中国海洋统计年鉴

CHINA MARINE STATISTICAL YEARBOOK 2007

国家海洋局

Edited by
State Oceanic Administration,
People's Republic of China

海洋出版社
China Ocean Press

图书在版编目(CIP)数据

中国海洋统计年鉴.2007:汉英对照/国家海洋局编. —北京:海洋出版社,2008.2
ISBN 978-7-5027-7024-2

Ⅰ. 中… Ⅱ. 国… Ⅲ. 海洋-统计资料-中国-2007-年鉴-汉、英 Ⅳ.P7-66

中国版本图书馆 CIP 数据核字(2008)第 057481 号

责任编辑:王 溪
责任印制:刘志恒

海洋出版社 出版发行
http://www.oceanpress.com.cn
北京市海淀区大慧寺路 8 号 邮编:100081
国家海洋信息中心印刷厂印刷 新华书店发行所经销
2008 年 2 月第 1 版 2008 年 2 月第 1 次印刷
开本:787mm×1092mm 1/16 印张:18.5
字数:439 千字 印数:1~1000 册
定价:168.00 元
发行部:(010)62147016 邮购部:(010)68038093 总编室:(010)62114335

《中国海洋统计年鉴》编委会

陈伟杰　　中国海洋石油有限公司
胡红江　　中国盐业总公司
王华俊　　中国有色金属工业协会
来　鹤　　辽宁海洋与渔业厅
张俊杰　　河北国土资源厅
孙连友　　天津海洋局
鲁小兵　　山东海洋与渔业厅
唐庆序　　江苏海洋与渔业局
张惠荣　　上海市海洋局
刘向东　　浙江海洋与渔业局
陈泽銮　　福建海洋与渔业局
文　斌　　广东海洋与渔业局
田凤鸣　　广西国土资源厅
黄良胜　　海南海洋与渔业厅
董艳涛　　大连市海洋与渔业局
赵泽斌　　青岛市海洋与渔业局
吴建义　　宁波市海洋与渔业局
周鲁闽　　厦门市海洋与渔业局
梁俊乾　　深圳市农林渔业局

《中国海洋统计年鉴》编辑部

刘燕伟	中国海洋石油有限公司
杜丽华	中国盐业总公司
孙秀敏	中国有色金属工业协会
魏　南	辽宁海洋与渔业厅
生金栋	河北国土资源厅
陶　钢	天津海洋局
史更新	山东海洋与渔业厅
唐雪峰	江苏海洋与渔业局
管永华	上海海洋局
应　晓	浙江海洋与渔业局
李锦松	福建海洋与渔业局
利红兵	广东海洋与渔业局
覃立团	广西国土资源厅
林　云	海南海洋与渔业厅
刘国伟	大连市海洋与渔业局
王子江	青岛市海洋与渔业局
忻佩忠	宁波市海洋与渔业局
林瑞才	厦门市海洋与渔业局
罗小霞	深圳市农林渔业局

执行编辑：张原华

英文校订：林宝法

Chen Weijie	China National Offshore Oil Corporation Limited
Hu Hongjiang	China National Salt Industry Corporation
Wang Huajun	China Nonferrous Metals Industry Association
Lai He	Department of Ocean and Fisheries of Liaoning Province
Zhang Junjie	Hebei Province Department of Land and Resources
Sun Lianyou	Tianjin Ocean Administration
Lu Xiaobing	Department of Oceanic and Fishery of Shandong Province
Tang Qingxu	Jiangsu Provincial Ocean and Fisheries Bureau
Zhang Huirong	Shanghai Municipal Ocean Bureau
Liu Xiangdong	Zhejiang Provincial Ocean and Fisheries Bureau
Chen Zeluan	Fujian Provincial Ocean and Fisheries Bureau
Wen Bin	Guangdong Provincial Oceanic and Fishery Administration
Tian Fengming	Department of Land and Resources of Guangxi
Huang Liangsheng	Department of Marine and Fishery of Hainan Province
Dong Yantao	Dalian Ocean and Fishery Administration
Zhao Zebin	Qingdao Ocean and Fishery Administration
Wu Jianyi	Ningbo Ocean and Fishery Bureau
Zhou Lumin	Oceans and Fisheries Bureau of Xiamen
Liang Junqian	Shenzhen Municipal Administration of Agriculture, Forestry and Fishery

Executive Editor: Zhang Yuanhua

English Proof-reader: Lin Baofa

编者说明

一、《中国海洋统计年鉴》2007年版是一部全面反映2006年中华人民共和国海洋经济发展和海洋管理服务情况的资料性年鉴，全书为中英文对照。

二、本年鉴的统计资料范围为人们在海洋和沿海地区开发、管理、利用海洋资源和空间，发展海洋经济的生产和活动以及沿海地区的社会经济概况。地域范围为沿海地区、沿海城市和沿海地带，其排列顺序按《沿海行政区域分类与代码》(HY/T 094-2006)的顺序排列。

三、年鉴资料内容包括综合资料、海洋经济核算、主要海洋产业活动、主要海洋产业生产能力、涉海就业、海洋科学技术、海洋教育、海洋环境保护、海洋行政管理及公益服务、全国及沿海社会经济、部分世界海洋经济统计资料等十一部分。

四、本年鉴根据《海洋统计报表制度》(国统制[2007]29号)和《海洋生产总值核算制度》(国统制[2006]21号)，资料主要来源于沿海省、自治区、直辖市统计局、海洋厅(局)以及20个有关涉海部、局、总公司。

五、本年鉴中除特殊说明外，所有价值量指标均为当年价。年鉴中每部分后附有主要海洋统计指标解释，对主要海洋统计指标的含义、统计范围和统计方法做了简要说明。统计数据中的其他说明置于表的下面。有续表的资料，如有注释均置于最后一张表的下面。

六、年鉴表格中符号使用说明："…"表示数据不足本表最小计算单位数；"空格"表示该项统计指标数据不详或无该项数据；"#"表示其中的主要项；其他符号如"*"或"①"等表示本表后面有注释。

七、由于数字精确度的原因，四舍五入后部分分项值之和与合计值有微小差异。

八、本年鉴资料国内部分未包括香港、澳门特别行政区和台湾省数据。

九、《中国海洋统计年鉴》在编汇过程中，得到了各有关单位的大力支持，我们在此表示衷心的感谢。本年鉴中如有疏漏和不妥之处，敬请读者批评指正。

《中国海洋统计年鉴》编辑部

Introduction

Ⅰ.***China Marine Statistical Yearbook (2007)*** is a data almanac which reflects in an all-round way the development of marine economy, marine management and service in the People's Republic of China in 2006, and it is a Chinese-English bilingual edition.

Ⅱ.The Yearbook's statistics cover the production and activities in the marine and coastal areas in relation to the development, management and utilization of marine resources and space, and the development of marine social economy. The regions covered are the coastal regions, coastal cities and coastal zones with coastlines, which are arranged in order according to the Coastal Administrative Areas Classification and Codes(HY/T 094-2006).

Ⅲ.The data in the yearbook consist of 11 sections, namely, integrated data, marine economic accounting, major marine industrial activities, production capacity of major marine industries, ocean-related employment, marine science and technology, marine education, marine environmental protection, marine administration and public-good service, national and coastal socioeconomy, part of the world's marine economic statistics data.

Ⅳ.The Yearbook is based on the Marine Statistics Report System (Guotongzhi [2007] No. 29) and the Ocean Gross Product Accounting System(Guotongzhi [2006] No. 21), its data mainly come from the statistical bureaus and the oceanic administrations of the coastal provinces, autonomous regions, and municipalities directly under the Central Government as well as the 20 ocean-related ministries, bureaus and general corporations concerned.

Ⅴ.Unless otherwise specified in the Yearbook, all the value indicators are given at the current price. Before each section are attached explanatory notes to the major marine statistical indicators, giving a brief explanation for the meaning, statistical range and statistical methods of the major marine statistical indicators. Other notes to the statistical data are listed below the tables. For the data with continued tables, annotations, if any, are put below the last table.

Ⅵ.The usage of symbols in the tables: "…" indicates the statistics smaller than the minimum calculation unit in the table; "Blank" indicates that the data of the statistical index are unknown for the time being or that there shouldn't be any data; "#" indicates the major items of the table; Other symbols, such as "*"or "①", indicate "see footnotes below".

Ⅶ.For reasons of digital accuracy, there is small difference between the sum of values of some subterms after having been rounded off and the total values.

Ⅷ.The domestic part of the Yearbook does not include that of Hong Kong Special Administrative Region, Macau Special Administrative Region and Taiwan Province.

Ⅸ.In the course of editing China Marine Statistical Yearbook, we enjoyed energetic support from the various departments concerned and we hereby extend our heartfelt thanks to them. Criticisms and comments are welcome from readers on any of the oversights and inappropriateness .

Editorial Department of
China Marine Statistical Yearbook

目　次
CONTENTS

3 主要海洋产业活动

Major Marine Industrial Activities

5 涉海就业

Ocean-Related Employment

6 海洋科学技术

Marine Science and Technology

8 海洋环境保护

Marine Environmental Protection

9 海洋行政管理及公益服务

Marine Administration and Public-Good Service

2006年我国海洋经济发展综述

2006年，是我国国民经济和社会发展第十一个五年规划实施的第一年，在党中央、国务院的正确领导下，沿海各级海洋行政主管部门坚持科学发展观，切实履行海洋管理职能，继续贯彻落实《全国海洋经济发展规划纲要》，不断推进海洋经济的持续健康发展。全国主要海洋产业继续保持稳步增长态势，海洋经济总量实现了新突破，海洋经济在国民经济和社会发展中的地位日益突出。

一、全国海洋经济发展概况

2006年，全国海洋生产总值突破了两万亿，达到21 220.3亿元，比上年增长16.3%，占国内生产总值和沿海地区生产总值比重分别为10.1%和15.7%；全国涉海就业人员2 960.3万人，占全国就业人员总数的3.9%，与“十五”期初相比增长了40.4%，对沿海地区乃至全国扩大就业起到了积极的作用，有力地促进了国家就业政策的贯彻实施。

“十五”期间，我国海洋经济继续呈现出稳健增长态势。全国海洋生产总值年均增长速度14.2%(除特别注明外，增长率均按可比价计算)，高出同期国内生产总值年均增长速度4.7个百分点。2005年，海洋生产总值占同期国内生产总值比重9.6%，比“十五”期初高出0.9个百分点。海洋经济快速发展的同时，有力地拉动了沿海地区就业持续增长，“十五”期末我国涉海就业人员2 780.8万人，比“十五”期初增加约670万个就业岗位。

二、主要海洋产业发展情况

2006年，全国主要海洋产业增加值8 792.5亿元，占全国海洋生产总值的比重达41.4%。“十五”期间，主要海洋产业保持稳步增长态势，海洋渔业、海洋交通运输业、滨海旅游业等海洋支柱产业继续保持领先优势，占主要海洋产业比重超过五分之四；海洋生物医药业、海水利用业等新兴海洋产业发展迅速，年均增长速度超过20%。

海洋第一产业 2006年，海洋渔业克服了台风、赤潮等灾害的影响，实现增加值1 708.1亿元，比上年增长6.6%。海水增养殖稳步发展，海水养殖产量1 445.64万吨，比上年增长4.39%；海洋捕捞有所回落，海洋捕捞产量1 442.04万吨，比上年减少11.26万吨，其中远洋捕捞产量为109.07万吨，比上年减少24.16%。

海洋第二产业 2006年，海洋第二产业稳步增长。其中，海洋船舶工业持续快速发展，新承接船舶订单及手持船舶订单与去年相比均有大幅增长，全年实现增加值为380.6亿元，比上年增长38.8%；海洋化工业快速增长，实现增加值184.6亿元，比上年增长20.0%；海洋电力业在国家政策的激励下，海洋能利用项目相继开工建设，全年实现增加值4.4亿元，比上年增长22.3%；海水利用业在《海水利用专项规划》的引导下稳步推进，全国实现增加值3.4亿元，比上年增长7.7%；海洋油气业继续保持增长态势，实现增加值668.9亿元，比上年增长3.8%，增速回落较大，海洋原油产量3 239.9万吨，比上年增长2.1%，占全国原油产量比重略有上升。

海洋第三产业 2006年，海洋第三产业快速增长。其中，海洋交通运输业实现增加值2 842.1亿元，比上年增长16.0%，全年主要沿海港口集装箱吞吐量80 969.0万吨，比上年增长43.6%，首次跃居世界第二；滨海旅游业增加值2 619.6亿元，比上年增长29.0%，接待入境旅游者3 439.89万人次，比上年增长13.1%。

三、区域海洋经济发展情况

2006 年，我国区域海洋经济继续保持良好的发展态势，环渤海经济区、长江三角洲和珠江三角洲海洋生产总值分别为 7 619.3 亿元、7 131.7 亿元和 4 113.9 亿元，分别占全国海洋生产总值的 35.9%、33.6% 和 19.4%。

“十五”期间，沿海地区加快海洋经济发展，努力培育海洋产业集群优势，环渤海经济区、长江三角洲经济区、珠江三角洲经济区已成为我国海洋经济密集程度较高的区域。“十五”期末，环渤海经济区海洋生产总值 6 263.8 亿元，年均增速 20.2%（按现价计算）；长江三角洲经济区海洋生产总值 5 837.5 亿元，年均增速 16.0%（按现价计算）；珠江三角洲经济区海洋生产总值 3 597.8 亿元，年均增速 14.8%（按现价计算）。

四、海洋科研教育情况

2006 年，海洋科研教育继续以服务海洋事业为发展方向，海洋科研项目为 6 593 项，发表海洋科技论文 8 492 篇，出版海洋科技著作 110 种，申请受理海洋科技专利 567 件；海洋专业博士、硕士研究生招生人数比上年增加 90 人，在校生人数为 6 075 人，毕业生人数为 1 270 人，比上年增加 264 人。

“十五”期间，海洋科学研究取得了较大的进展，累计海洋科研项目近 18 000 项，海洋高等教育培养了一批海洋专业人才，培养的博士、硕士毕业人数累计 3 086 人，为我国海洋事业发展提供了坚实的科技和人才保障。

五、海洋环境保护情况

2006 年，全海域未达到清洁海域水质标准的面积 14.9 万平方公里，比上年增长6.9%；沿海地区工业废水排放量1 426 096万吨，比上年减少1.9%，工业废水排放达标率为92.9%，比上年降低了3个百分点；全国沿海18个海洋生态监控区，有14个海洋生态监控区处于亚健康或不健康状态，其中不健康状态海洋生态监控区占生态监控区总数的22.2%；沿海地区赤潮灾害发生9次，受灾面积10 555平方公里。

“十五”期间，我国海域总体污染状况尚未好转，近岸海域污染形势依然严峻，海洋环境保护任务艰巨。沿海地带工业废水排放中直接排放入海量占三分之一。

六、海洋行政管理与公益服务情况

2006年，全国颁发海域使用权证书8 759本，确权海域面积22.7万公顷；签发疏浚物海洋倾倒许可证466份，倾倒区面积352.6平方公里；全年共实施海洋执法检查次数45 088次，发现违法行为3 101起；海洋数值预报服务共943 085次，海洋监测获得数据量7 522.8万个，海洋调查项目836个。

“十五”期间，我国海洋管理工作逐步规范有序，截至2005年末全国确权海域面积951 978.1公顷，发放海域使用权证书31 979本；签发疏浚物海洋倾倒许可证2 749份，倾倒区面积333.0平方公里；海洋执法检查次数累计75 965次，发现违法行为9 268起。海洋公益服务日趋完善，截至2005年末沿海地区海滨观测台站835个，海洋预报服务444 511次，海洋监测获得数据量26 450.8万组，海洋调查项目2 930个，全局室（馆）存纸介质档案74 954卷（册），实际接收存档卫星遥感数据量达440 208.3GB。

Summary of China´s Marine Economic Development in 2006

2006 is the first year for implementing the China's 11th Five-Year Plan for National Economy and Social Development. Under the correct leadership of the Central Party Committee and the State Council, the coastal competent marine administrative departments at all levels adhere to the concept of scientific development, earnestly fulfill the functions of marine administration, continue to carry out the National Programme for Marine Economic Development, and constantly promote the sustained and healthy development of marine economy. The major marine industries throughout China continue to keep a posture of steady growth, a new breakthrough has been made in the marine economic aggregates, and the position of marine economy in the national economy and social development has become increasingly prominent.

I .Survey of the National Marine Economic Development.

In 2006, the gross output value of China's major marine industries has topped 2 trillion yuan (RMB), amounting to 2.12203 trillion yuan, up 16.3% from the previous year, its proportion in the GDP and the Gross Regional Product being 10.1% and 15.7% respectively; The number of people employed by the ocean-related sectors throughout the country reaches 29.603 million, accounting for 3.9% of the national total employed personnel, registering an increase of 40.4% as compared with the early period of the 11th Five-Year Plan, which has played a positive role in expanding the employment in the coastal region and the country as a whole, and given a strong impetus to the implementation of the nation's employment policies.

During the 10th Five-Year Plan, China's marine economy continued to present a pattern of steady growth. The annual average growth rate of the national gross ocean product was 14.2 % (Unless otherwise specified, the growth rate is calculated at the comparable price.), rising by 4.7 percentage points against the annual average growth rate of GDP in the same period. In 2005, the proportion of the gross ocean product in the GDP in the same period was 9.6%, 0.9 percentage points higher than in the early 10th Five-Year Plan period. While it was rapidly developing, the marine economy had vigorously spurred the sustained increase of employment in the coastal region. At the end of the 11th Five-Year Plan period, the number of people employed by the ocean related sectors reached 27.808 million throughout China, about 6.70 million more job opportunities than that in the early 10th Five-Year Plan period.

Ⅱ.Development Situation of the Major Marine Industries.

In 2006, the national added value of major marine industries was 879.25 billion yuan, occupying a proportion of 41.4% in the national gross ocean product. During the10th Five-Year Plan, the major marine industries maintained a posture of steady growth, and the pillar marine industries such as marine fishery, marine communications and transportation, and coastal tourism, etc. continued to hold the lead, their proportion in the major marine industries exceeding 4/5; the new emerging marine industries like marine biomedicine and seawater utilization industry developed rapidly with an annual increase rate of over 20%.

Primary marine industry In 2006, having overcome the effects of such disasters as typhoon and red tide, marine fishery effects an added value of 170.81 billion yuan, up 6.6% from the pervious year. Seawater culture and stock enhancement develops steadily, the production of mariculture amounts to 14.4564 million tons, 4.39% up from the previous year; Slightly falling, the yield of marine fishing reaches 14.4204 million tons, recording a reduction of 0.1126 million tons as against that of the previous year, among which, the yield of ocean-going fishing is 1.0907 million tons, down 24.16% from the previous year.

Secondary marine industry In 2006, the secondary marine industry grows steadily, of which, the marine shipping industry develops in a rapid and sustained way, the newly undertaken ship orders and the ship orders in hand have seen an increase by a large margin as compared with the previous year, accomplishing an added value of 38.06 billion yuan for the whole year, up 38.8% from the previous year; Marine chemical industry rapidly grows, effecting an added value of 18.46 billion yuan, increasing by 20.0% as against that in the previous year; For the marine electric power industry, spurred by the state policies, the projects of ocean energies utilization have started construction one after another, realizing an added value of 0.44 billion yuan, up 22.3% from the previous year; The seawater utilization industry steadily advances under the guidance of the Special Project Plan for Seawater Utilization effects an added value 0.34 billion yuan, up 7.7% from the previous year; The offshore oil and gas industry continues to maintain a posture of growth, and effects an added value of 66.89 billion yuan, up 3.8% from the previous year, but with a great drop in the rate of growth, and the offshore crude oil production accounts to 32.399 million tons, rising by 2.1% as compared with the previous year, and its proportion in the national total crude oil production slightly rises.

Tertiary marine industry In 2006, the tertiary marine rapidly grows. The marine communications and transportation industry yields an added value of 284.21 billion yuan, up 16.0% from the previous year and the container handling capacity of the major coastal ports reaches 809.69 million tons, up 43.6% from the previous year, vaulting into the second place in the world for the first time; The added value of coastal tourism amounts to 261.96 billion yuan, up 29.0% from the previous year, and the number of inbound tourists received is 34.3989 million person-times, up 13.1% from the previous year.

Ⅲ.Development Condition of Regional Marine Economy

In 2006, China's regional marine economy continues to maintain a good situation of development. The gross ocean products of the Round-the-Bohai Economic Zone, Changjiang River Delta and Zhujiang River Delta are 761.93 billion yuan, 713.17 billion and 411.39 billion yuan respectively, accounting for 35.9%, 33.6% and 19.4% of the national gross ocean product.

During the 10th Five-Year Plan, the coastal regions speeded up the marine economic development, striving to nurture the grouping superiority of marine industries. The Round-the-Bohai Economic Zone, Changjiang River Delta Economic Zone and Zhujiang River Delta Economic Zone have now become areas with a higher level of concentration of marine economy in China. At the end of the 10th Five-Year Plan period, gross ocean product of the Round-the-Bohai Economic Zone amounted to 626.38 billion yuan, with an annual growth rate of 20.2%(calculated at the current price), that of the Changjiang River Delta Economic Zone 583.75 billion yuan, with an annual growth rate of 16.0% (calculated at the current price) and that of the Zhujiang River Delta Economic Zone 359.78 billion yuan, with an annual growth rate of 14.8% (calculated at the current price).

Ⅳ.Marine Science, Technology and Education

In 2006, marine science, technology and education continues its developing orientation of serving the marine cause. The number of marine scientific research projects reaches 6 593, a total of 8 492 marine scientific and technological papers and 110 kinds of marine scientific and technological works are published, and 567 applications for marine scientific and technological patents are dealt with; The number of postgraduates enrolled for doctors' and masters' degrees increases by 90 as compared with that in the previous year, the number of undergraduates is 6 075 and that of graduates 1 270, 264 more than that in the previous year.

During the 10th Five-Year Plan period, the marine scientific research made greater progress, with a total number of nearly 18 000 marine scientific research projects. Marine higher education trained a large number of marine specialized personnel and the number of doctor and master graduates in marine specialties totalled 3 086, thus providing solid guarantee of science & technology, and talents for the development of marine cause.

V.Marine Environmental Protection

In 2006, the whole sea area that is not up to the standard of clean sea area water quality reaches 149 000 km^2, increasing by 6.9% as compared with that in the previous year; The discharge amount of industrial waste water in the coastal region is 14.26096 billion tons, down 1.9% from the previous year, the up-to-standard rate of industrial waste water is 92.9%, dropping by 3 percentage points as against that in the precious year; Among the 18 marine ecological monitoring areas, 14 are in a sub-healthy or unhealthy state, of which the unhealthy marine ecological monitoring areas occupy 22.2% of the total marine ecological monitoring areas; Red tide disaster occurs nine times in the coastal region and the disaster-affected area amounts to 10 555 km^2.

During the 10th Five-Year Plan period, the overall condition of pollution in China's sea area was not improved; the pollution situation was still grave and the task of protecting the marine environment is formidable. In the coastal zone, the discharge amount of industrial waste water accounted for one third of the total.

VI.Marine Administration & Public Service

In 2006, a total of 8 759 certificates for the sea area use right are issued throughout the country, the sea area with the ownership of patent rights reaching 227 000 hm^2; 466 permits for oceanic dumping of dredged material are signed and issued, the area of the dumping zone amounting to 352.6 km^2; Throughout the year the marine inspections for law enforcement are carried out on 450 88 occasions in which 3 101 cases of unlawful practice were found; Marine forecast service is provided on

943 085 occasions, the date quantity obtained from marine monitoring is 752 280 00 and the number of marine survey projects reaches 836.

During the 10th Five-Year Plan period, China's marine management work has gradually become standardized and orderly. By the end of 2005, the sea area with the ownership of patent rights amounted to 951 978.1 hm^2 with 31 979 certificates of sea area use rights issued; A total of 2 749 certificates for oceanic dumping of dredgers material were signed and issued, with the area of the dumping zone amounting to 333.0 km^2; Marine inspection for law enforcement was carried out on 75 965 occasions in which 9 268 cases of unlawful practice were found. The marine public service tended to be perfect day by day, and by the end of 2005, the number of coastal observation stations reached 835, marine forecast service was provided on 444 511 occasions, the quantity of date obtained from marine monitoring was 0.264 508 billion sets, the number of marine survey projects amounted to 2 930, the archives throughout the State Oceanic Administration had a collection of 74 954 volumes of paper files and the quantity of data from the satellite remote-sensing which were actually received and kept in the archives reached 440 208.3GB.

图1 全国海洋生产总值及三次产业构成
China's Gross Ocean Product and Three Industries Composition

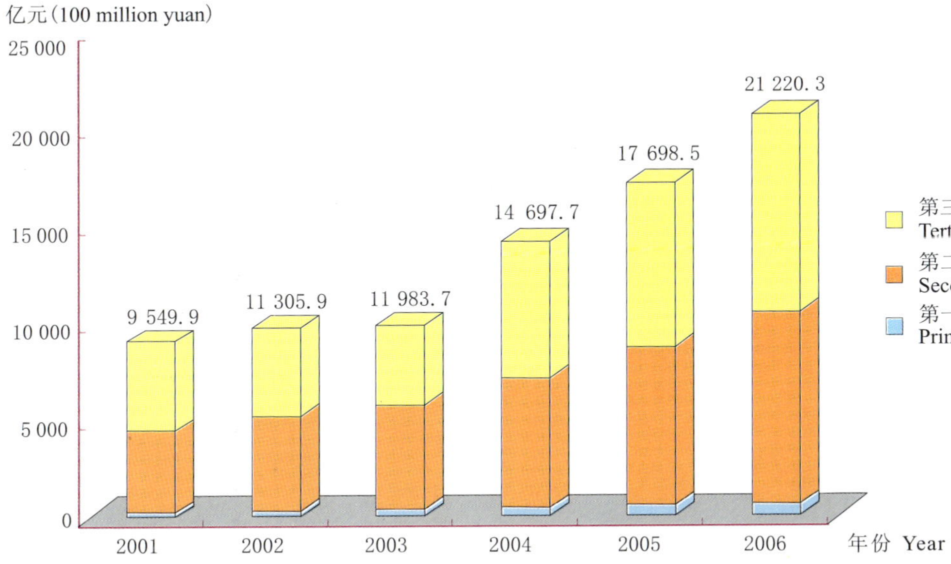

图2 2006年全国主要海洋产业增加值构成
Composition of Added Values of the Major Marine Industries in China in 2006

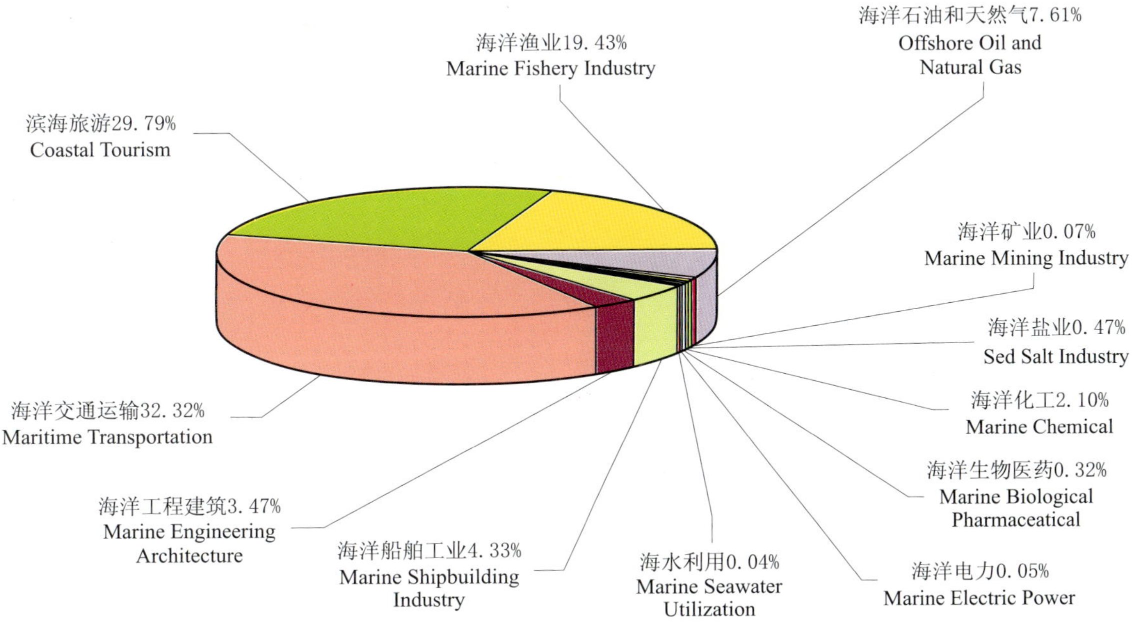

图3 2006年沿海地区海洋生产总值
Gross Ocean Product by Coastal Regions in 2006

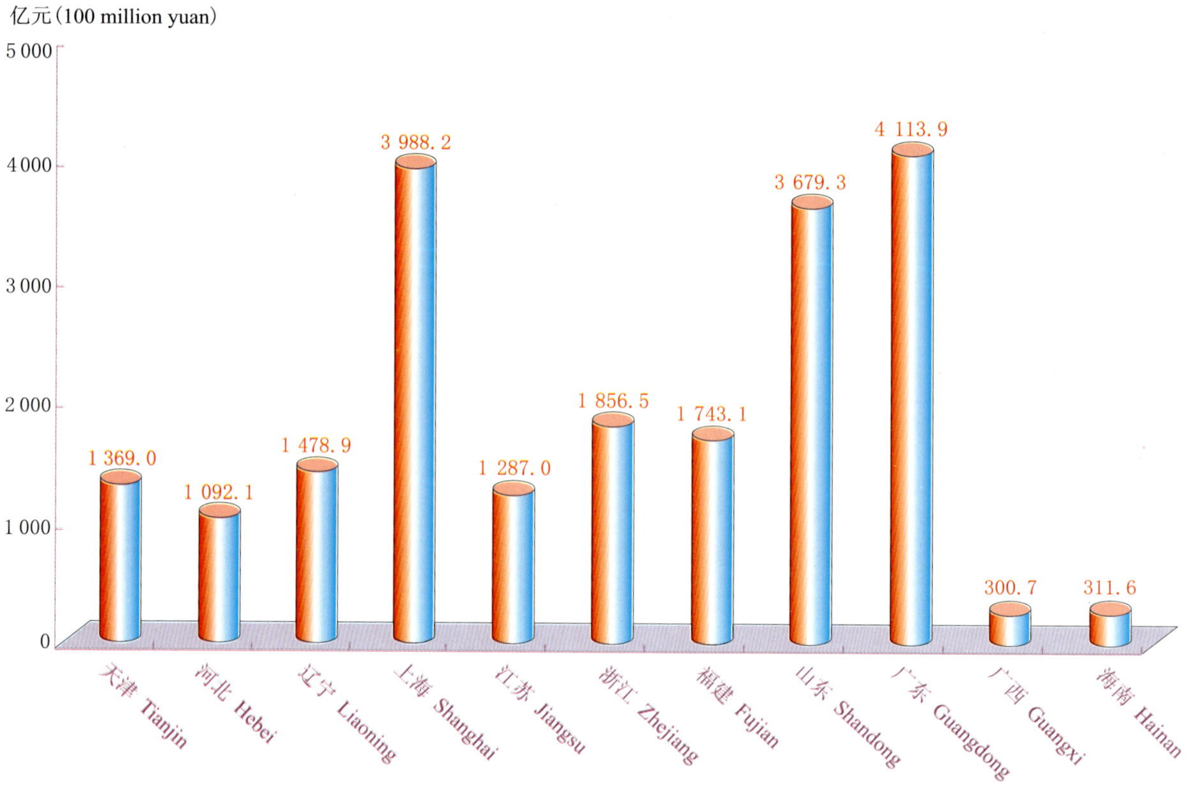

图4 全国海洋捕捞养殖产量
National Marine Fishing and Mariculture Production

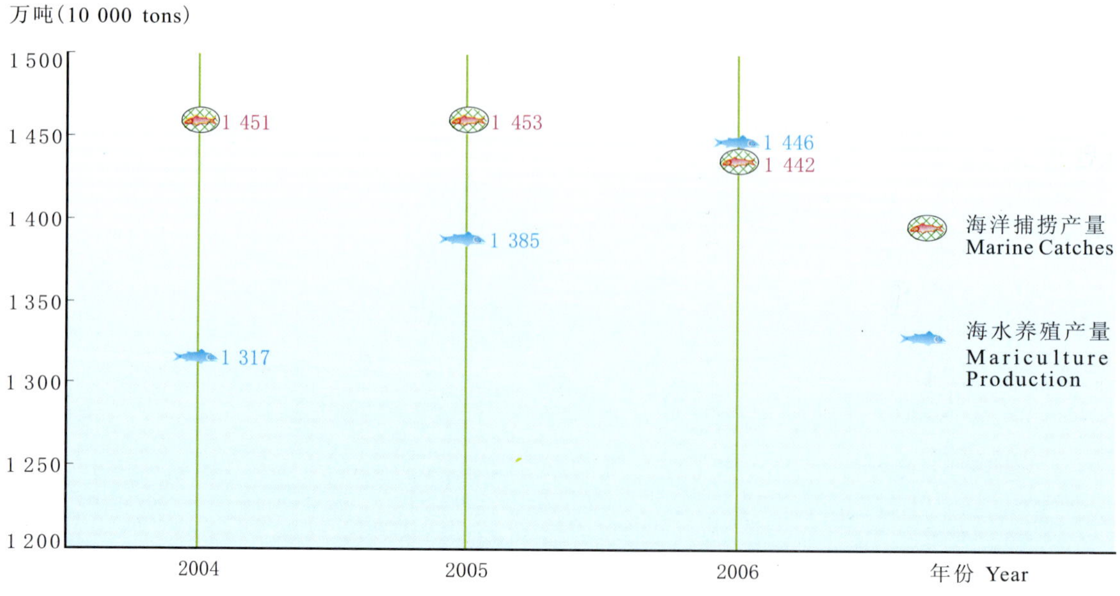

图5 2006年沿海地区海洋捕捞养殖产量
Marine Fishing and Mariculture Production by Coastal Regions in 2006

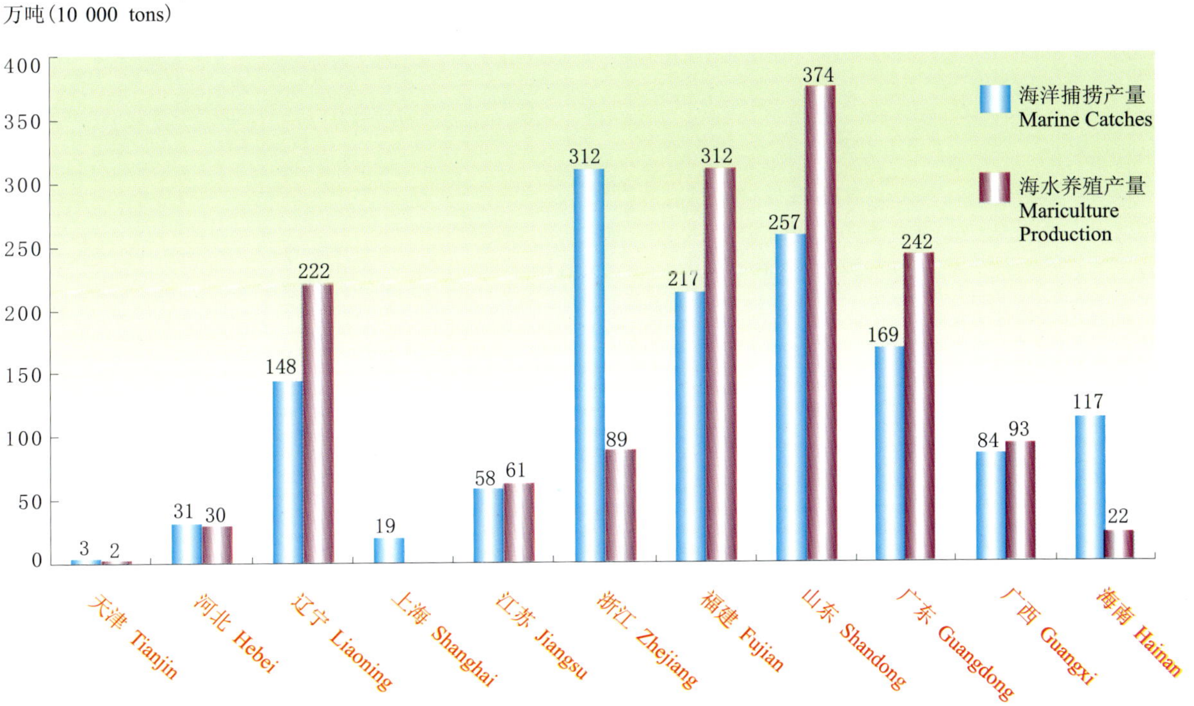

图6 全国海洋石油和天然气产量
National Output of Offshore Oil and Natural Gas

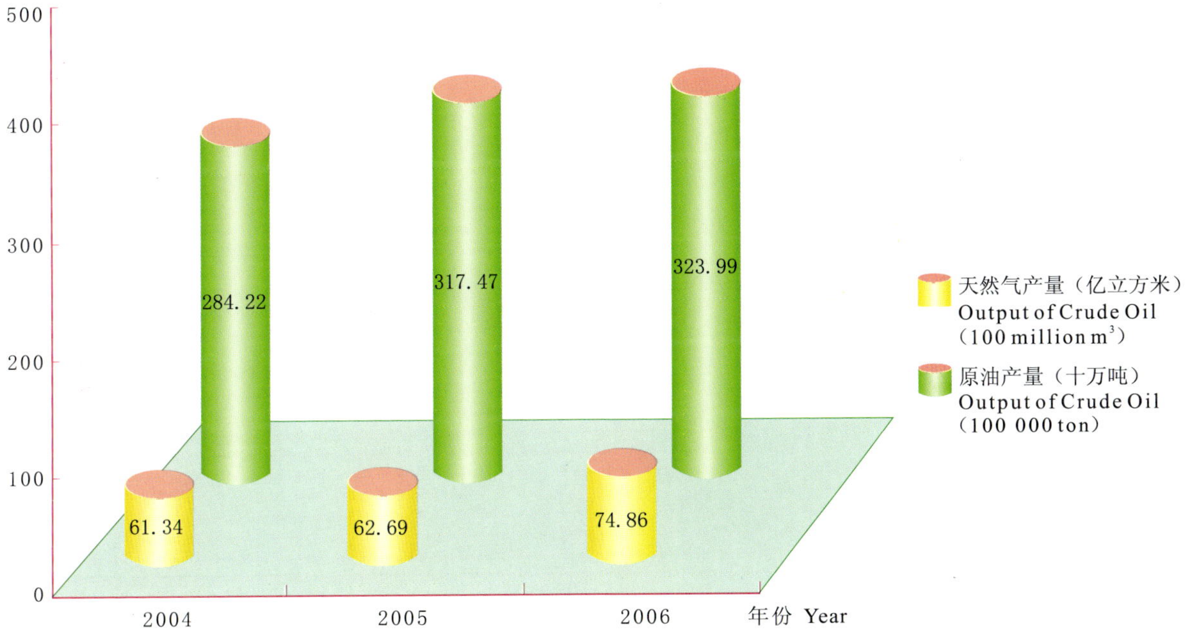

图7 海洋原油产量占全国原油产量比重
Production of Offshore Crude Oil Production in the National Total

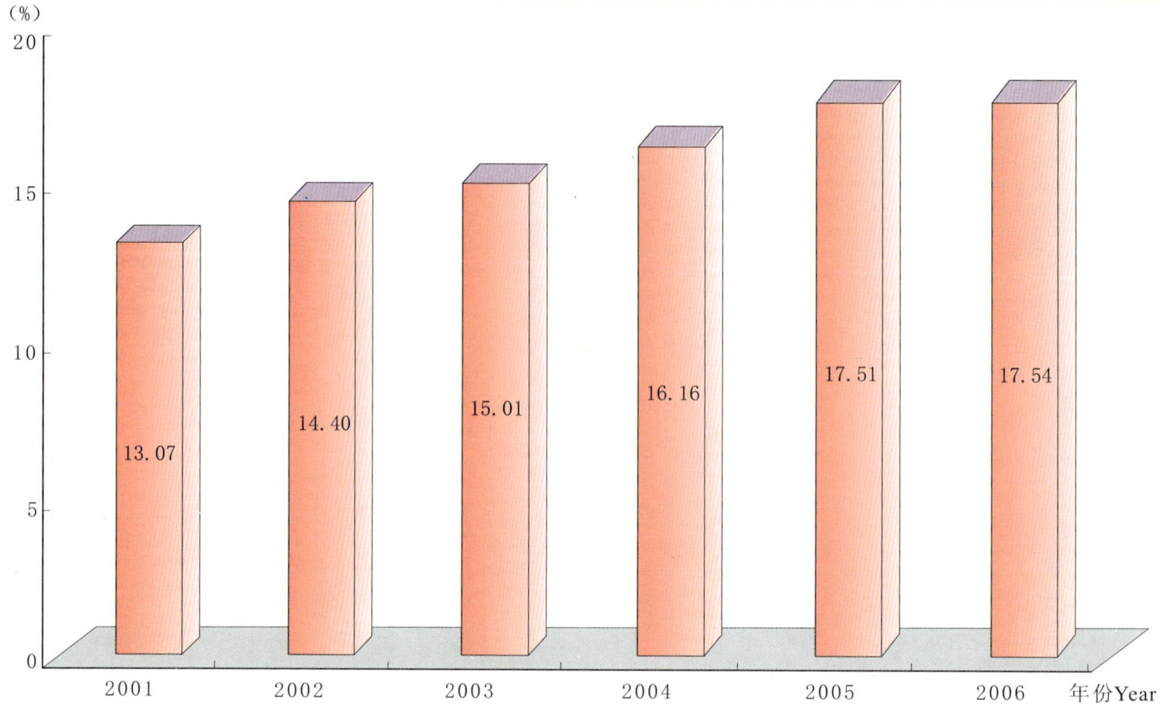

图8 2006年海滨砂矿产量
Production of Marine Mining Industry in 2006

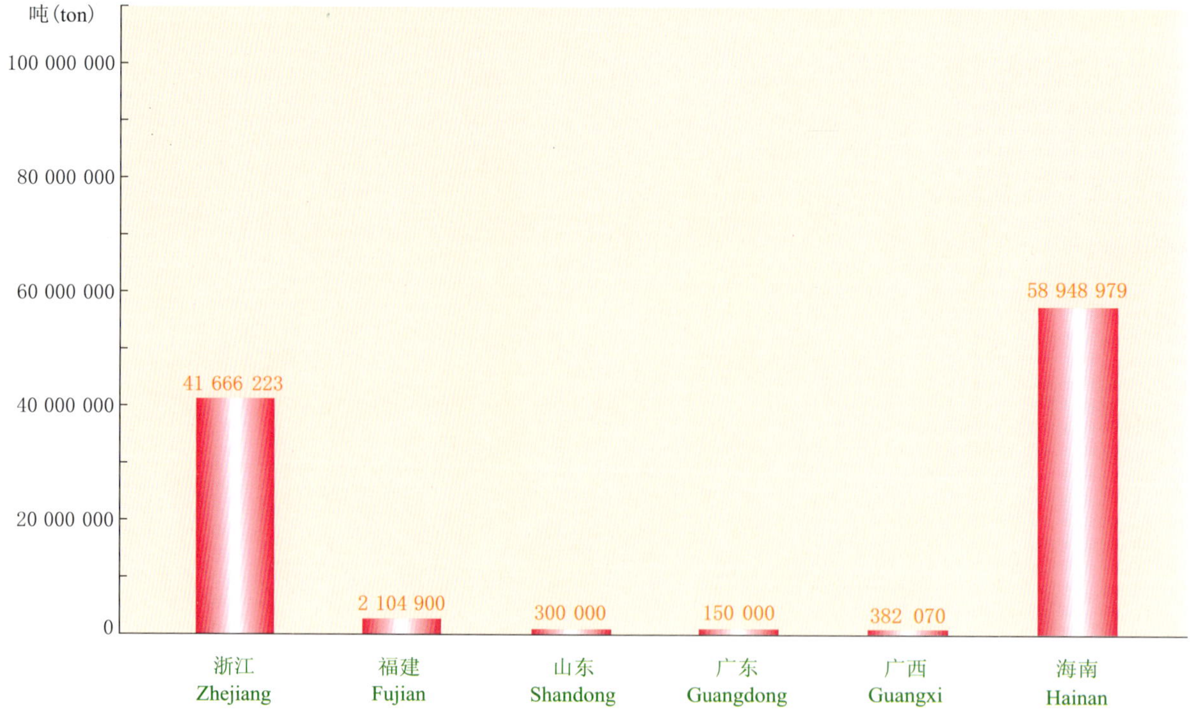

图9　2006年沿海地区海盐产量
Sea Salt Production by Coastal Regions in 2006

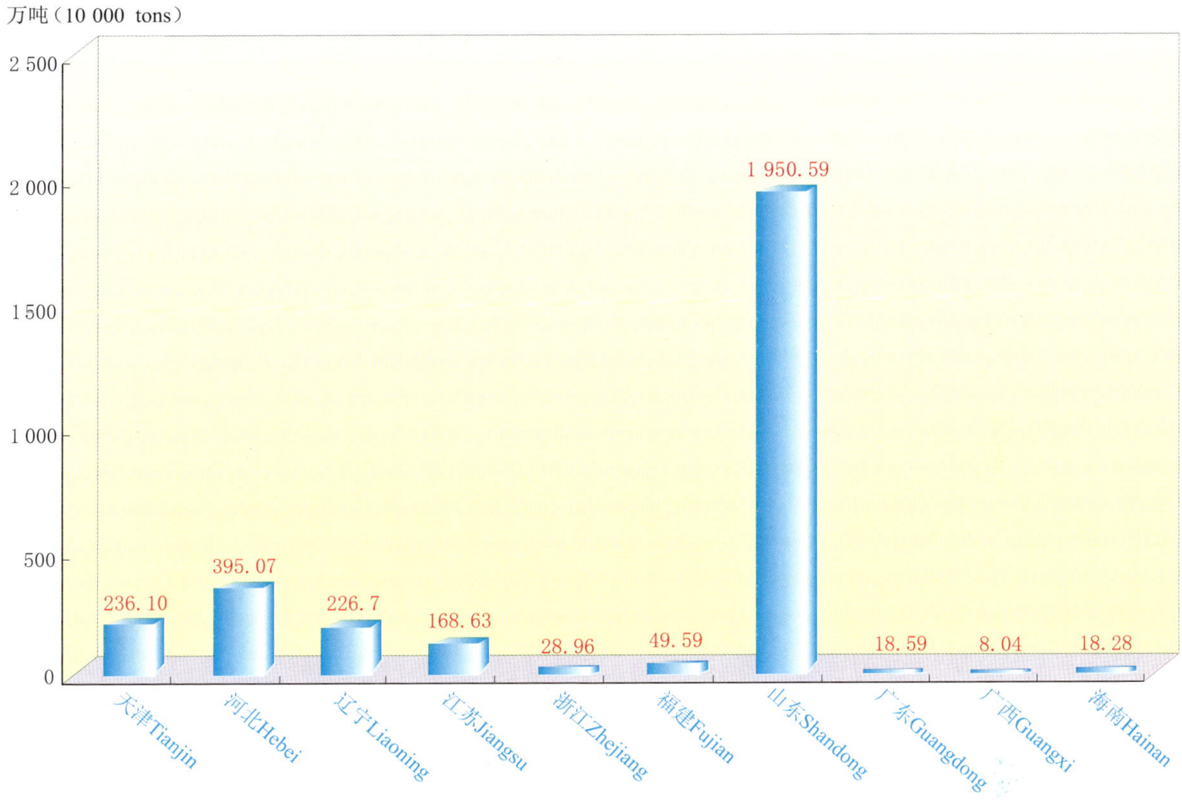

图10　2006年沿海地区造船完工量
Completed Number of Ships Built by Coastal Regions in 2006

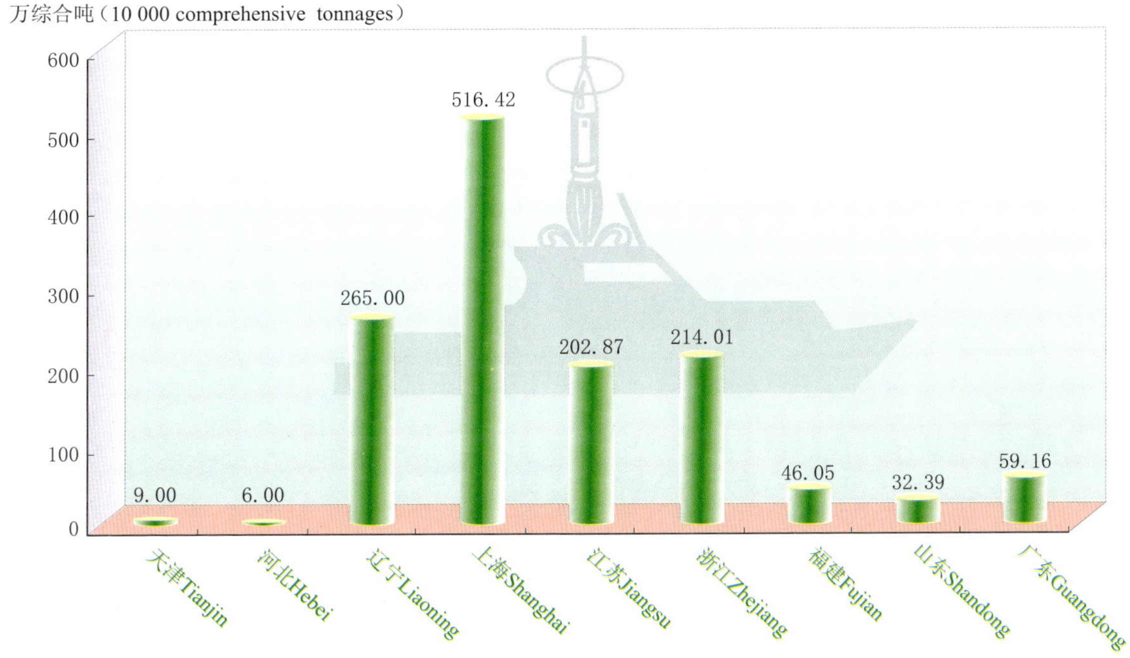

图11 2006年沿海地区海洋货物周转量
Cargo Turnover Volume by Coastal Regions in 2006

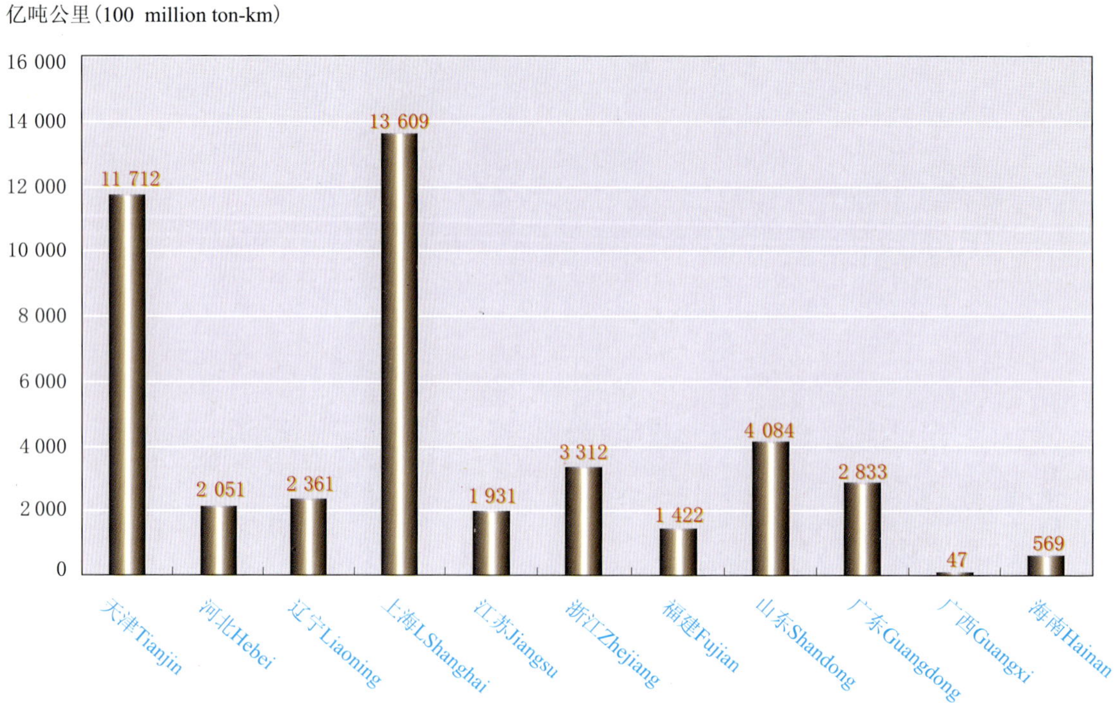

图12 2006年沿海地区主要沿海港口国际标准集装箱吞吐量
International Standardized Containers Handled at Principal Seaports by Coastal Regions in 2006

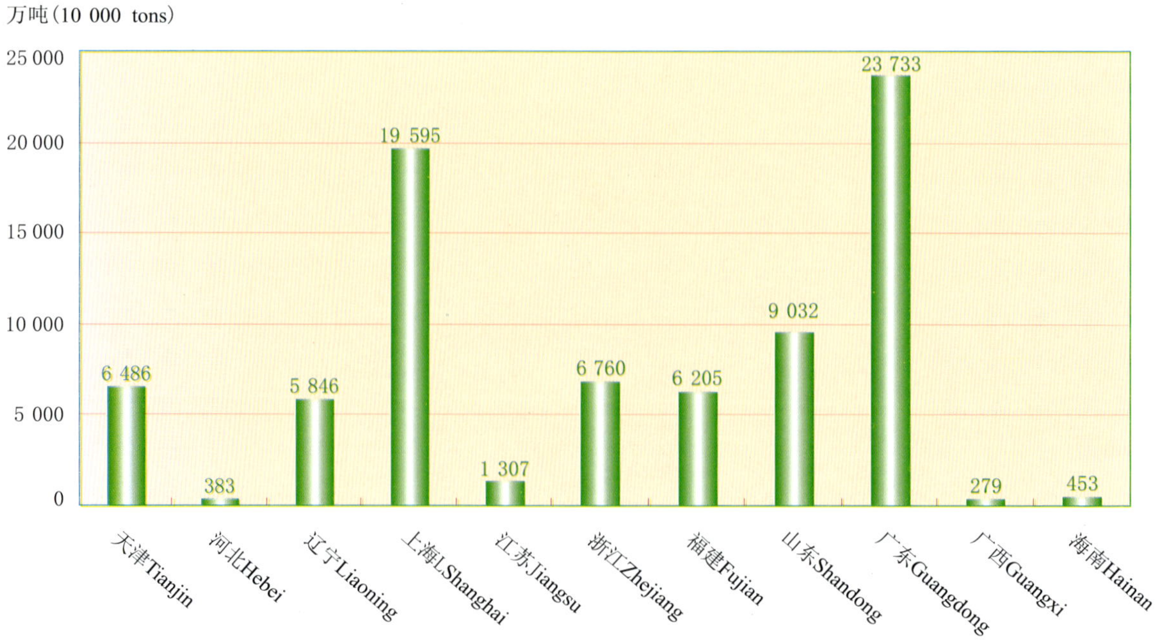

图13　2006年沿海地区滨海旅游收入
Earning from Coastal Tourism by Coastal Regions in 2006

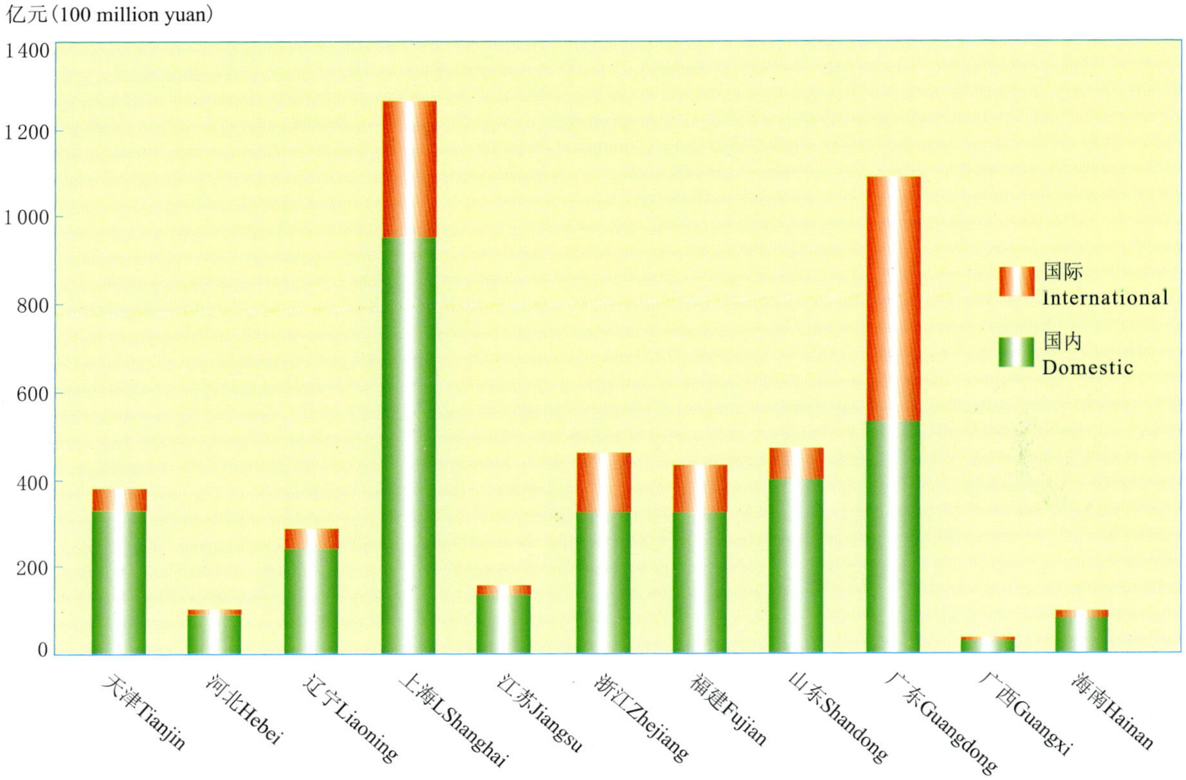

图14　沿海城市国际旅游收入和接待入境旅游者人数
Earnings from International Tourism and Number of Tourists Received by Coastal Cities

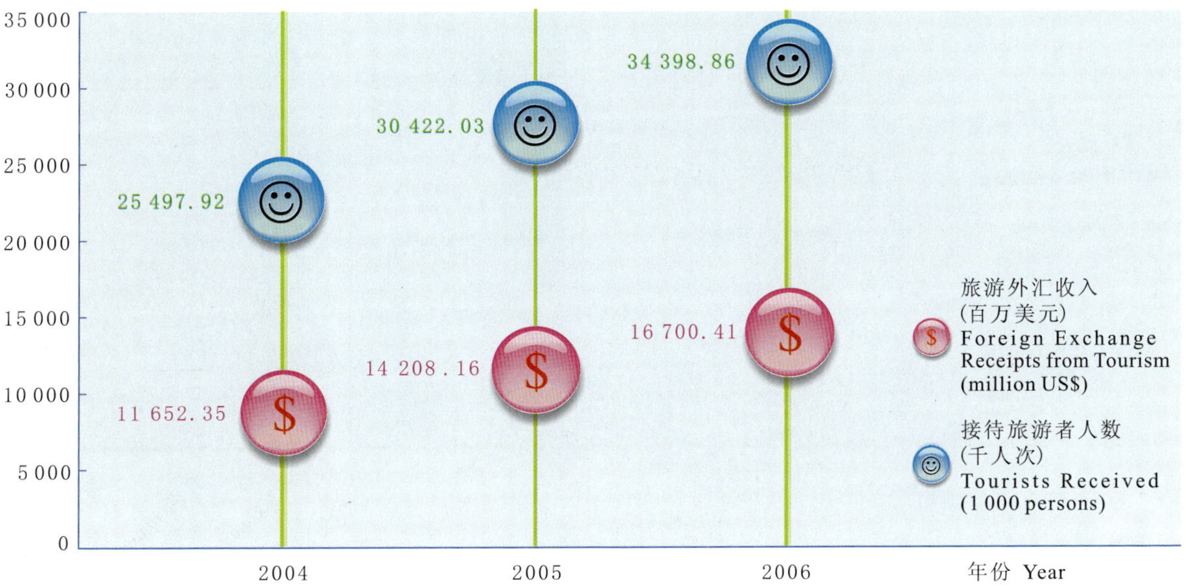

图15 2006年沿海地区海洋经济贡献
Marine Economic Contributions of the Coastal Regions in 2006

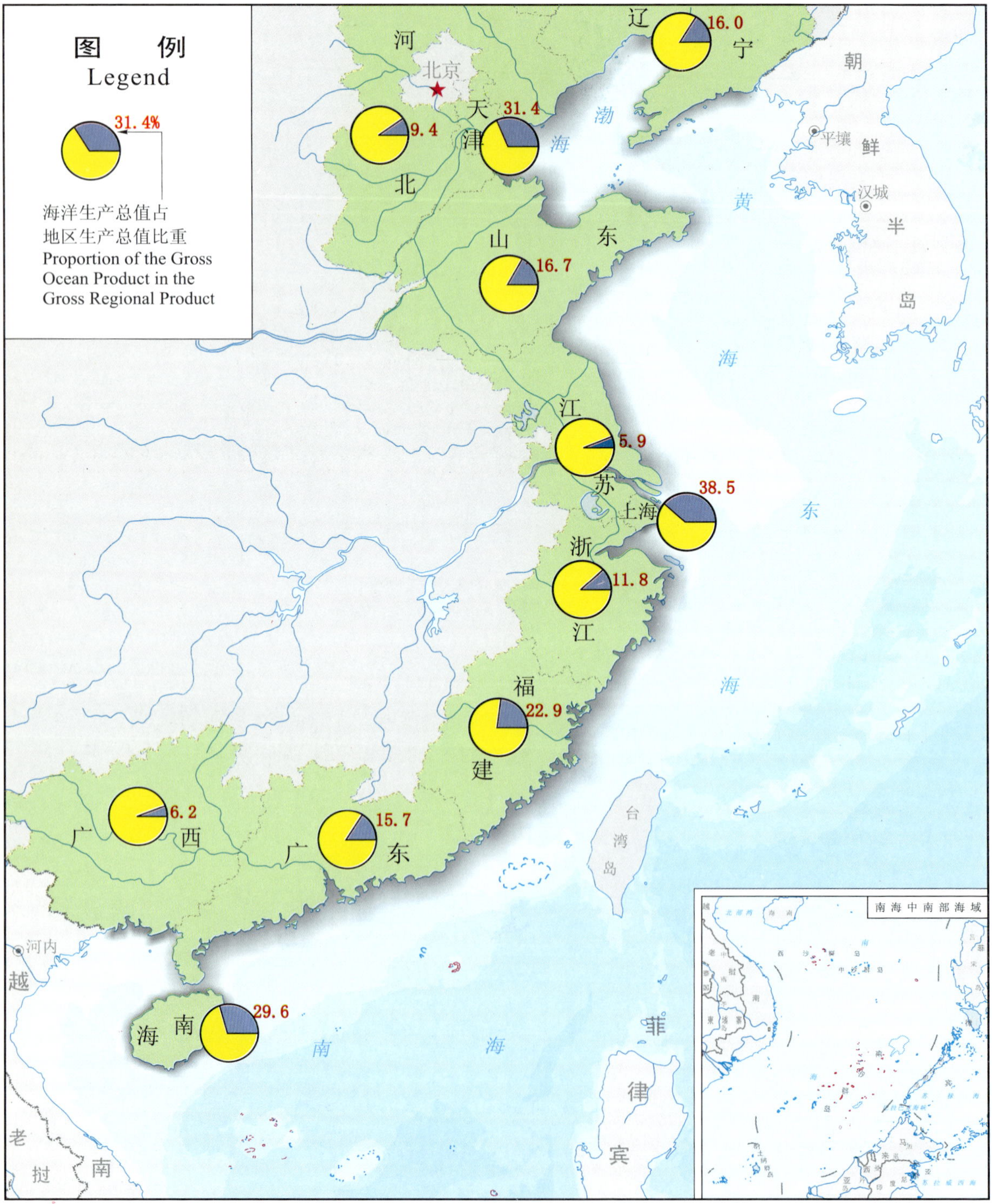

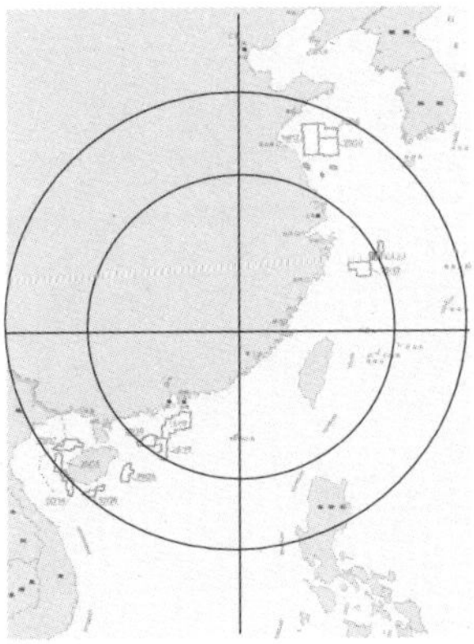

1

综合资料

INTEGRATED DATA

1-1 沿海地区行政区划（2006年）
Administrative Division of Coastal Regions, 2006

单位：个 Unit: number

沿海地区 Coastal Region	沿海城市 Coastal City	沿海地带 Coastal County (District)			
		合　计 Total	县 County	县级市 County-level city	区 District
合　计 **Total**	**53**	**242**	**65**	**57**	**120**
天　津 Tianjin	1	3			3
河　北 Hebei	3	11	6	1	4
辽　宁 Liaoning	6	22	4	7	11
上　海 Shanghai	1	6	1		5
江　苏 Jiangsu	3	15	9	5	1
浙　江 Zhejiang	7	37	11	10	16
福　建 Fujian	6	34	11	8	15
山　东 Shandong	7	37	6	14	17
广　东 Guangdong	14	56	11	6	39
广　西 Guangxi	3	8	1	1	6
海　南 Hainan	2	13	5	5	3

注：沿海地带中未包括广东省的东莞、中山和海南省的三亚。

Note: Dongguan, Zhongshan of Guangdong province and Sanya of Hainan province are not included in the "Coastal County".

1-2 沿海行政区划一览表（2006年）
Table of Administrative Division of Coastal Regions, 2006

沿海地区 Coastal Region	地区代码 Zip Code	沿海城市 Coastal City	地区代码 Zip Code	沿海地带 Coastal County (District)	地区代码 Zip Code
天 津 Tianjin	120000			塘沽区Tanggu Qu	120107
				汉沽区Hangu Qu	120108
				大港区Dagang Qu	120109
河 北 Hebei	130000	唐山 Tangshan	130200	丰南区Fengnan Qu	130207
				滦南县Luannan Xian	130224
				乐亭县Leting Xian	130225
				唐海县Tanghai Xian	130230
		秦皇岛 Qinhuangdao	130300	海港区Haigang Qu	130302
				山海关区Shanhaiguan Qu	130303
				北戴河区Beidaihe Qu	130304
				昌黎县Changli Xian	130322
				抚宁县Funing Xian	130323
		沧州 Cangzhou	130900	海兴县Haixing Xian	130924
				黄骅市Huanghua Shi	130983
辽 宁 Liaoning	210000	大连 Dalian	210200	中山区Zhongshan Qu	210202
				西岗区Xigang Qu	210203
				沙河口区Shahekou Qu	210204
				甘井子区Ganjingzi Qu	210211
				旅顺口区Lüshunkou Qu	210212
				金州区Jinzhou Qu	210213
				长海县Changhai Xian	210224
				瓦房店市Wafangdian Shi	210281
				普兰店市Pulandian Shi	210282
				庄河市Zhuanghe Shi	210283
		丹东 Dandong	210600	东港市Donggang Shi	210681
		锦州 Jinzhou	210700	凌海市Linghai Shi	210781
		营口 Yingkou	210800	西市区Xishi Qu	210803
				鲅鱼圈区Bayuquan Qu	210804
				老边区Laobian Qu	210811
				盖州市Gaizhou Shi	210881
		盘锦 Panjin	211100	大洼县Dawa Xian	211121
				盘山县Panshan Xian	211122

续表 continued

沿海地区 Coastal Region	地区代码 Zip Code	沿海城市 Coastal City	地区代码 Zip Code	沿海地带 Coastal County (District)	地区代码 Zip Code
		葫芦岛 Huludao	211400	连山区Lianshan Qu	211402
				龙港区Longgang Qu	211403
				绥中县Suizhong Xian	211421
				兴城市Xingcheng Shi	211481
上海 Shanghai	310000			宝山区Baoshan Qu	310113
				浦东新区Pudong Xinqu	310115
				金山区Jinshan Qu	310116
				南汇区Nanhui Qu	310119
				奉贤区Fengxian Qu	310120
				崇明县Chongming Xian	310230
江苏 Jiangsu	320000	南通 Nantong	320600	海安县Hai'an Xian	320621
				如东县Rudong Xian	320623
				启东市Qidong Shi	320681
				通州市Tongzhou Shi	320683
				海门市Haimen Shi	320684
		连云港 Lianyungang	320700	连云区Lianyun Qu	320703
				赣榆县Ganyu Xian	320721
				东海县Donghai Xian	320722
				灌云县Guanyun Xian	320723
				灌南县Guannan Xian	320724
		盐城 Yancheng	320900	响水县Xiangshui Xian	320921
				滨海县Binhai Xian	320922
				射阳县Sheyang Xian	320924
				东台市Dongtai Shi	320981
				大丰市Dafeng Shi	320982
浙江 Zhejiang	330000	杭州 Hangzhou	330100	滨江区Binjiang Qu	330108
				萧山区Xiaoshan Qu	330109
				余杭区Yuhang Qu	330110
		宁波 Ningbo	330200	海曙区Haishu Qu	330203
				江东区Jiangdong Qu	330204
				江北区Jiangbei Qu	330205
				北仑区Beilun Qu	330206
				镇海区Zhenhai Qu	330211
				鄞州区Yinzhou Qu	330212
				象山县Xiangshan Xian	330225

续表 continued

沿海地区 Coastal Region	地区代码 Zip Code	沿海城市 Coastal City	地区代码 Zip Code	沿海地带 Coastal County (District)	地区代码 Zip Code
				宁海县Ninghai Xian	330226
				余姚市Yuyao Shi	330281
				慈溪市Cixi Shi	330282
				奉化市Fenghua Shi	330283
		温州 Wenzhou	330300	鹿城区Lucheng Qu	330302
				龙湾区Longwan Qu	330303
				瓯海区Ouhai Qu	330304
				洞头县Dongtou Xian	330322
				平阳县Pingyang Xian	330326
				苍南县Cangnan Xian	330327
				瑞安市Rui'an Shi	330381
				乐清市Yueqing Shi	330382
		嘉兴 Jiaxing	330400	海盐县Haiyan Xian	330424
				海宁市Haining Shi	330481
				平湖市Pinghu Shi	330482
		绍兴 Shaoxing	330600	绍兴县Shaoxing Xian	330621
				上虞市Shangyu Shi	330682
		舟山 Zhoushan	330900	定海区Dinghai Qu	330902
				普陀区Putuo Qu	330903
				岱山县Daishan Xian	330921
				嵊泗县Shengsi Xian	330922
		台州 Taizhou	331000	椒江区Jiaojiang Qu	331002
				路桥区Luqiao Qu	331004
				玉环县Yuhuan Xian	331021
				三门县Sanmen Xian	331022
				温岭市Wenling Shi	331081
				临海市Linhai Shi	331082
福建 Fujian	350000	福州 Fuzhou	350100	马尾区Mawei Qu	350105
				连江县Lianjiang Xian	350122
				罗源县Luoyuan Xian	350123
				平潭县Pingtan Xian	350128
				福清市Fuqing Shi	350181
				长乐市Changle Shi	350182
		厦门 Xiamen	350200	思明区Siming Qu	350203
				海沧区Haicang Qu	350205

续表 continued

沿海地区 Coastal Region	地区代码 Zip Code	沿海城市 Coastal City	地区代码 Zip Code	沿海地带 Coastal County (District)	地区代码 Zip Code
				湖里区Huli Qu	350206
				集美区Jimei Qu	350211
				同安区Tong'an Qu	350212
				翔安区Xiang'an Qu	350213
		莆田 Putian	350300	城厢区Chengxiang Qu	350302
				涵江区Hanjiang Qu	350303
				荔城区Licheng Qu	350304
				秀屿区Xiuyu Qu	350305
				仙游县Xianyou Xian	350322
		泉州 Quanzhou	350500	丰泽区Fengze Qu	350503
				洛江区Luojiang Qu	350504
				泉港区Quangang Qu	350505
				惠安县Hui'an Xian	350521
				金门县Jinmen Xian	350527
				石狮市Shishi Shi	350581
				晋江市Jinjiang Shi	350582
				南安市Nan'an Shi	350583
		漳州 Zhangzhou	350600	云霄县Yunxiao Xian	350622
				漳浦县Zhangpu Xian	350623
				诏安县Zhao'an Xian	350624
				东山县Dongshan Xian	350626
				龙海市Longhai Shi	350681
		宁德 Ningde	350900	蕉城区Jiaocheng Qu	350902
				霞浦县Xiapu Xian	350921
				福安市Fu'an Shi	350981
				福鼎市Fuding Shi	350982
山东 Shandong	370000	青岛 Qingdao	370200	市南区Shinan Qu	370202
				市北区Shibei Qu	370203
				四方区Sifang Qu	370205
				黄岛区Huangdao Qu	370211
				崂山区Laoshan Qu	370212
				李沧区Licang Qu	370213
				城阳区Chengyang Qu	370214
				胶州市Jiaozhou Shi	370281
				即墨市Jimo Shi	370282
				胶南市Jiaonan Shi	370284

续表 continued

沿海地区 Coastal Region	地区代码 Zip Code	沿海城市 Coastal City	地区代码 Zip Code	沿海地带 Coastal County (District)	地区代码 Zip Code
		东营 Dongying	370500	东营区Dongying Qu	370502
				河口区Hekou Qu	370503
				垦利县Kenli Xian	370521
				利津县Lijin Xian	370522
				广饶县Guangrao Xian	370523
		烟台 Yantai	370600	芝罘区Zhifu Qu	370602
				福山区Fushan Qu	370611
				牟平区Muping Qu	370612
				莱山区Laishan Qu	370613
				长岛县Changdao Xian	370634
				龙口市Longkou Shi	370681
				莱阳市Laiyang Shi	370682
				莱州市Laizhou Shi	370683
				蓬莱市Penglai Shi	370684
				招远市Zhaoyuan Shi	370685
				海阳市Haiyang Shi	370687
		潍坊 Weifang	370700	寒亭区Hanting Qu	370703
				寿光市Shouguang Shi	370783
				昌邑市Changyi Shi	370786
		威海 Weihai	371000	环翠区Huancui Qu	371002
				文登市Wendeng Shi	371081
				荣成市Rongcheng Shi	371082
				乳山市Rushan Shi	371083
		日照 Rizhao	371100	东港区Donggang Qu	371102
				岚山区Lanshan Qu	371103
		滨州 Binzhou	371600	无棣县Wudi Xian	371623
				沾化县Zhanhua Xian	371624
广东 Guangdong	440000	广州 Guangzhou	440100	荔湾区Liwan Qu	440103
				越秀区Yuexiu Qu	440104
				海珠区Haizhu Qu	440105
				天河区Tianhe Qu	440106
				白云区Baiyun Qu	440111
				黄埔区Huangpu Qu	440112
				番禺区Panyu Qu	440113
				南沙区Nansha Qu	440115
				萝岗区Luogang QU	440116

续表 continued

沿海地区 Coastal Region	地区代码 Zip Code	沿海城市 Coastal City	地区代码 Zip Code	沿海地带 Coastal County (District)	地区代码 Zip Code
		深圳 Shenzhen	440300	罗湖区Luohu Qu	440303
				福田区Futian Qu	440304
				南山区Nanshan Qu	440305
				宝安区Bao'an Qu	440306
				龙岗区Longgang Qu	440307
				盐田区Yantian Qu	440308
		珠海 Zhuhai	440400	香洲区Xiangzhou Qu	440402
				斗门区Doumen Qu	440403
				金湾区Jinwan Qu	440404
		汕头 Shantou	440500	龙湖区Longhu Qu	440507
				金平区Jinping Qu	440511
				濠江区Haojiang Qu	440512
				潮阳区Chaoyang Qu	440513
				潮南区Chaonan Qu	440514
				澄海区Chenghai Qu	440583
				南澳县Nan'ao Xian	440523
		江门 Jiangmen	440700	蓬江区Pengjiang Qu	440703
				江海区Jianghai Qu	440704
				新会区Xinhui Qu	440705
				台山市Taishan Shi	440781
				恩平市Enping Shi	440785
		湛江 Zhanjiang	440800	赤坎区Chikan Qu	440802
				霞山区Xiashan Qu	440803
				坡头区Potou Qu	440804
				麻章区Mazhang Qu	440811
				遂溪县Suixi Xian	440823
				徐闻县Xuwen Xian	440825
				廉江市Lianjiang Shi	440881
				雷州市Leizhou Shi	440082
				吴川市Wuchuan Shi	440083
		茂名 Maoming	440900	茂南区Maonan Qu	440902
				茂港区Maogang Qu	440903
				电白县Dianbai Xian	440923
		惠州 Huizhou	441300	惠城区Huicheng Qu	441302
				惠阳区Huiyang Qu	441303
				惠东县Huidong Xian	441323

续表 continued

沿海地区 Coastal Region	地区代码 Zip Code	沿海城市 Coastal City	地区代码 Zip Code	沿海地带 Coastal County (District)	地区代码 Zip Code
		汕尾 Shanwei	441500	城　区Chengqu	441502
				海丰县Haifeng Xian	441521
				陆丰市Lufeng Shi	441581
		阳江 Yangjiang	441700	江城区Jiangcheng Qu	441702
				阳西县Yangxi Xian	441721
				阳东县Yangdong Xian	441723
		东莞 Dongguan	441900		
		中山 Zhongshan	442000		
		潮州 Chaozhou	445100	湘桥区Xiangqiao Qu	445102
				饶平县Raoping Xian	445122
		揭阳 Jieyang	445200	榕城区Rongcheng Qu	445202
				揭东县Jiedong Xian	445221
				惠来县Huilai Xian	445224
广西 Guangxi	450000	北海 Beihai	450500	海城区Haicheng Qu	450502
				银海区Yinhai Qu	450503
				铁山港区Tieshangang Qu	450512
				合浦县Hepu Xian	450521
		防城港 Fangchenggang	450600	港口区Gangkou Qu	450602
				防城区Fangcheng Qu	450603
				东兴市Dongxing Shi	450681
		钦州 Qinzhou	450700	钦南区Qinnan Qu	450702
海南 Hainan	460000	海口 Haikou	460100	秀英区Xiuying Qu	460105
				龙华区Longhua Qu	460106
				美兰区Meilan Qu	460108
		三亚 Sanya	460200		
				琼海市 Qionghai Shi	469002
				儋州市 Danzhou Shi	469003
				文昌市 Wenchang Shi	469005
				万宁市 Wanning Shi	469006
				东方市 Dongfang Shi	469007
				澄迈县Chengmai Xian	469023
				临高县Lingao Xian	469024
				昌江黎族自治县Changjian	469026
				乐东黎族自治县Ledong Li	469027
				陵水黎族自治县Lingshui I	469028

1-3 海洋自然地理
Marine Physical Geography

指　　标		Item	指标值 Data
海洋平均深度	(米)	Average Depth of Sea (m)	961
海洋最大深度	(米)	Maximum Depth of Sea (m)	5 559
岸线总长度	(公里)	Length of Coastline (km)	32 000
大陆岸线长度		Mainland Shore	18 000
岛屿岸线长度		Island Shore	14 000
岛屿个数	(个)	Number of Islands	5 400
岛屿面积	(万平方公里)	Area of Islands (10 000 km^2)	3. 87

注：数据来源于《2007年中国统计年鉴》（表1-4同）。岛屿面积未包括香港、澳门特别行政区和台湾省。

Note: The data come from "China Statistical Yearbook 2007"（same as Table 1-4）. Island area does not include that of Hong Speical Administrative Region,Macao Speical Administrative Region and Taiwan Province.

1-4 海区海域面积
Sea Area

自然海区名称 Natural Sea Area		海域总面积 (千公顷) Sea Area (1 000 hm^2)	平均深度 (米) Average Depth (m)	最大深度 (米) Maximum Depth (m)
合 计	**Total**	**472 700**		
渤　海	Bohai Sea	7 700	18	70
黄　海	Huanghai Sea	38 000	44	140
东　海	Donghai Sea	77 000	370	2 719
南　海	Nanhai Sea	350 000	1 212	5 559

1-5 海洋自然资源
Ocean Natural Resources

指　　标	Item	指标值 Data
海洋能源理论蕴藏量　（亿千瓦）	Theoretical Sea-energy Reserves (100 million kW)	6. 3
海水可养殖面积　（万公顷）	Cultivatable Area in Marine Areas (10 000 hm^2)	260. 01
#已养殖面积	Cultivated Area	109. 49
浅海滩涂可养殖面积　（万公顷）	Cultivable Area in Shallow Sea and Sea-beaches (10 000 hm^2)	242. 00
#已养殖面积	Cultivated Area	89. 37

注：数据来源于《2007年中国统计年鉴》。
Note: The data come from "China Statistical Yearbook 2007".

1-6 海区海洋石油储量（2006年）

Offshore Oil Reserves in the Sea Area, 2006

自然海区名称 Natural Sea Area	海洋石油(万吨) Offshore Oil (10 000 tons)	
	累计探明技术可采储量 Proven Technically Recoverable Reserves in the Aggregate	剩余技术可采储量 Surplus Technically Recoverable Reserves
合 计 Total	**59 953**	**34 636**
渤 海 Bohai Sea	34 398	26 235
黄 海 Huanghai Sea		
东 海 Donghai Sea	745	449
南 海 Nanhai Sea	24 810	7 952

注：数据来源于《2006年全国油气矿产储量通报》。

Note: The data come from "Journal on the National Oil Gas and Mineral Resources Reserves in 2006".

1-7 沿海地区水资源基本情况（2006年）
Water Resources by Coastal Regions, 2006

地区 Region	水资源总量（亿立方米） Total Amount of Water Resources (100 million m^3)	地表水资源量 Surface Water Resources	地下水资源量 Groundwater Resources	地表水与地下水资源重复量 Duplicated Measurement Between Surface Water and Groundwater	人均水资源量（立方米/人） Per Capita Water Resources (m^3/person)
全国总计 National Total	**25 330.1**	**24 358.1**	**7 642.9**	**6 670.8**	**1 932.1**
天津 Tianjin	10.1	6.6	4.4	0.9	95.5
河北 Hebei	107.3	42.1	94.3	29.0	156.1
辽宁 Liaoning	261.4	233.9	91.6	64.2	615.5
上海 Shanghai	27.6	27.6	9.5	9.5	153.9
江苏 Jiangsu	404.4	314.7	110.7	21.0	538.3
浙江 Zhejiang	903.6	889.4	204.4	190.2	1 829.5
福建 Fujian	1 623.5	1 622.3	454.6	453.4	4 577.7
山东 Shandong	199.3	109.6	146.3	56.5	214.8
广东 Guangdong	2 216.2	2 206.5	487.0	477.4	2 396.1
广西 Guangxi	1 881.1	1 881.1	453.2	453.2	4 011.3
海南 Hainan	227.6	225.1	56.5	54.0	2 735.4

注：数据来源于《2007年中国统计年鉴》。

Note: The data come from "China Statistical Yearbook 2007".

1-8 沿海地区湿地面积
Area of Wetlands by Coastal Regions

地 区 Region	湿地面积 (千公顷) Area of Wetlands (1 000 hm^2)	天然湿地 Natural Wetlands	近岸及海岸 Coasts and Seashores	湿地面积占国土面积比重 (%) Proportion of Wetlands in Total Area of Territory (%)
全国总计 National Total	**38 485.5**	**36 200.6**	**5 941.7**	**4.01**
天 津 Tianjin	171.8	133.7	58.1	15.0
河 北 Hebei	1 081.9	1 042.3	278.8	5.8
辽 宁 Liaoning	1 219.6	1 106.8	738.1	8.4
上 海 Shanghai	319.7	319.4	305.4	53.7
江 苏 Jiangsu	1 674.7	1 651.1	843.5	16.3
浙 江 Zhejiang	802.2	695.9	574.3	7.9
福 建 Fujian	443.0	421.2	370.6	3.7
山 东 Shandong	1 784.1	1 681.4	1 210.9	11.7
广 东 Guangdong	1 398.1	1 252.0	1 017.8	7.9
广 西 Guangxi	656.1	567.5	348.4	2.8
海 南 Hainan	311.5	256.6	190.0	9.1

注：数据来源于《2007年中国统计年鉴》。本表为中国首次湿地调查 (1995-2003)资料，不包括台湾省、香港和澳门特别行政区；湿地面积不包括水稻田湿地。

Note: The data come from "China Statistical Yearbook 2007".Data in the table are the figures of China First Wetlands Survey (1995-2003), excluding the wetlands of Taiwan province, Hong Kong SAR and Macao SAR. Area of wetlands excludes the wetland of paddyfield.

1-9 红树林各地类面积

Site Classification and Area of Sharpleaf Mangrove (Rhizophora Apiculata)

单位：公顷 Unit：hm^2

地 区 Region	红树林各地类总面积 Total Site Area of Sharpleaf Mangrove	现有面积 Established	未成林面积 Unestablished	宜林地面积 Suitable for Planting
全国总计 National Total	**82 757.2**	**22 024.9**	**1 884.1**	**58 848.2**
浙 江 Zhejiang	5 452.3	20.6	236.1	5 195.6
福 建 Fujian	13 410.1	615.1	286.4	12 508.6
广 东 Guangdong	32 325.9	9 084.0	981.3	22 260.6
广 西 Guangxi	18 029.2	8 374.9	380.3	9 274.0
海 南 Hainan	13 539.7	3 930.3		9 609.4

注：数据来源于《2007年中国统计年鉴》。本表数据为2002年全国红树林资源调查资料。

Note: The data come from "China Statistical Yearbook 2007".Data in the table are the figures of National Sharpleaf Mangrove Survey in 2002.

1-10 主要沿海城市气候基本情况（2006年）

Climate of Major Coastal Cities, 2006

城 市 City	年平均气温（摄氏度） Annual Average Temperature(℃)	年平均相对湿度（%） Annual Average Relative Humidity	全年降水量（毫米） Annual Precipitation (millimeter)	全年日照时数（小时） Annual Sunshine Hours (hour)
天 津 Tianjin	13.2	61	415.2	2 130.0
上 海 Shanghai	17.9	70	1 149.7	1 678.1
杭 州 Hangzhou	18.2	71	1 164.3	1 422.8
福 州 Fuzhou	20.8	72	1 904.7	1 564.3
广 州 Guangzhou	23.2	71	2 175.7	1 328.7
海 口 Haikou	25.4	78	1 547.7	1 848.8

注：数据来源于《2007年中国统计年鉴》。

Note: The data come from "China Statistical Yearbook 2007".

主要统计指标解释

1.沿海地区 即广义的沿海地区是指有海岸线(大陆岸线和岛屿岸线)的地区,按行政区划分为沿海省、自治区、直辖市。

2.沿海城市 是指有海岸线的直辖市和地级市(包括其下属的全部区、县和县级市)。

3.沿海地带 即狭义的沿海地区,是指有海岸线的县、县级市、区(包括直辖市和地级市的区)。

4.海洋 是海和洋的统称。洋为地球表面上相连接的广大咸水水体的主体部分。海为地球表面相连接的广大咸水水体被陆地、岛礁、半岛包围或分隔的边缘部分。

5.海水可养殖面积 指利用滩涂、浅海、港湾进行鱼、虾、蟹、贝、藻等海水经济动植物的人工养殖的水面面积。

6.水资源总量 指评价区内降水形成的地表和地下产水总量,即地表产流量与降水入渗补给地下水量之和,不包括过境水量。

7.地表水资源量 指评价区内河流、湖泊、冰川等地表水体中可以逐年更新的动态水量,即当地天然河川径流量。

8.地下水资源量 指评价区内降水和地表水对饱水岩土层的补给量,包括降水入渗补给量和河道、湖库、渠系、渠灌田间等地表水体的入渗补给量。

9.地表水与地下水资源重复量 指地表水和地下水相互转化的部分,即天然河川径流量中的地下水排泄量和地下水补给量中来源于地表水的入渗补给量。

10.湿地 指天然或人工、长久或暂时性的沼泽地、泥炭地或水域地带,包括静止或流动、淡水、半咸水、咸水体,低潮时水深不超过 6 米的水域以及海岸地带地区的珊瑚滩和海草床、滩涂、红树林、河口、河流、淡水沼泽、沼泽森林、湖泊、盐沼及盐湖。

11.红树林 指生长在热带、亚热带低能海岸潮间带上部,受周期性潮水浸淹,以红树植物为主体的常绿灌木或乔木组成的潮滩湿地木本生物群落。

12.气温 指空气的温度,我国一般以摄氏度(℃)为单位表示。气象观测的温度表是放在离地面约 1.5 米处通风良好的百叶箱里测量的,因此,通常说的气温指的是离地面 1.5 米处百叶箱中的温度。其统计计算方法为:

月平均气温是将全月各日的平均气温相加,除以该月的天数而得。

年平均气温是将 12 个月的月平均气温累加后除以 12 而得。

13.相对湿度 指空气中实际所含水蒸气密度和同温度下饱和水蒸气密度的百分比值。其统计方法与气温相同。

14.降水量 指从天空降落到地面的液态或固态(经融化后)水,未经蒸发、渗透、流失而在地面上积聚的深度。其统计计算方法为:

月降水量是将全月各日的降水量累加而得。

年降水量是将 12 个月的月降水量累加而得。

15.日照时数 指太阳实际照射地面的时间。其统计方法与降水量相同。

Explanatory Notes on Main Statistical Indicators

1.Coastal Region, i.e., the coastal region in a broad sense, refers to the regions with coastlines (continental and island coastlines), which are divided into the coastal provinces,autonomous regions and municipalities directly under the Central Government according to the administrative zoning.

2.Coastal City refers to the municipalities directly under the Central Government and the prefecture-level cities (including all the districts, counties and county-level cities under them).

3.Coastal Zone, i.e., the coastal region in a narrow sense, refers to the counties, county-level cities and districts with coastlines (including the districts under the municipalities directly under the Central Government and the prefecture-level districts).

4.Ocean is the general name for sea and ocean. Ocean refers to the main body of large salt water connected with the earth surface. Sea refers to the edge areas of the salt water on the earth surface that are compartmentalized or surrounded by land, island, reef or peninsula.

5.Marine Cultivatable Areas refer to water areas in beach, shallow sea and bays that are used to breed marine cash propagation, such as fish, shrimp, crab, shellfish, alga and so on.

6.Total Water Resources refers to the total volume of water resources measured as run-off for surface water from rainfall and recharge for groundwater in a given area, excluding transit water.

7.Surface Water Resources refers to the total renewable resources which exist in rivers, lakes, glaciers and other collectors from rainfall and are measured as run-off of rivers.

8.Groundwater Resources refers to the replenishment of aquifers with rainfall and surface water.

9.Duplicated Measurement between Surface Water and Groundwater refers to the exchange between surface water and groundwater, i.e. run-off of rivers includes some depletion into groundwater while groundwater includes some replenishment from surface water.

10.Wetlands refer to marshland and peat bog, whether natural or man-made, permanent or temporary; water covered areas, whether stagnant or flowing, with fresh or brackish-fresh or salty water that is less than 6 meters deep at low tide; as well as coral beach, weed beach, mud beach, mangrove, river outlet, rivers, fresh-water marshland, marshland forests, lakes, salty bog and salt lakes along the coastal areas.

11.Mangrove refers to evergreen woody plants or plant communities in tropical or sub-tropical zones which live between the sea and the land in areas which are inundated by tides.

12.Temperature refers to the air temperature. China uses centigrade as the unit. The thermometry used for weather observation is put in a breezy shutter, which is 1.5 meters high from the ground. Therefore, the commonly used temperature refers to the temperature in the breezy shutter 1.5 meters away from the ground. The calculation method is as follows:

Monthly average temperature is the summation of average daily temperature of one month divided by the actual days of that particular month.

Annual average temperature is the summation of monthly averages of a year divided by 12 months.

13.Relative Humidity refers to the ratio of actual water vapour pressure to the saturated water vapour density under the current temperature. The calculation method is the same as that of temperature.

14.Volume of Precipitation refers to the deepness of liquid state or solid state (thawed) water falling from the sky to the ground that has not evaporated, infiltrated or run off. The calculation method is as follows:

Monthly precipitation is the summation of daily precipitation of a month.

Annual precipitation is the summation of 12 months precipitation of a year.

15.Sunshine Hours refer to the actual hours of sun irradiating the earth. The calculation method is the same as that of the precipitation.

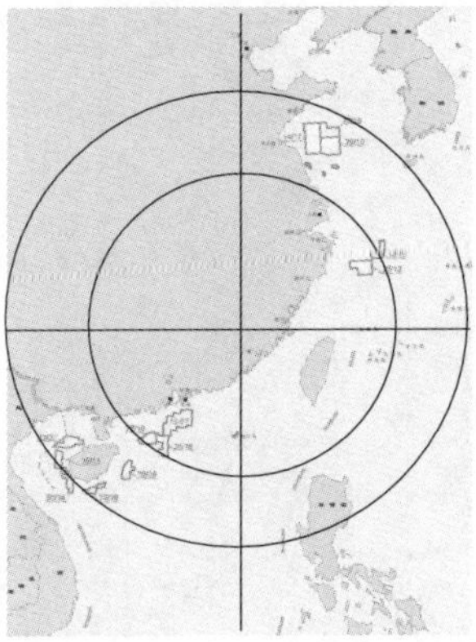

2

海 洋 经 济 核 算
Marine Economic Accounting

2-1 全国海洋生产总值

National Gross Ocean Product

年 份 Year	海洋生产总值（亿元） Gross Ocean Product (100 million yuan)				海洋生产总值占国内生产总值比重（%） Proportion of the Gross Ocean Product in GDP	海洋生产总值增长速度（%） Growth Rate of the Gross Ocean Product
		第一产业 Primary Industry	第二产业 Secondary Industry	第三产业 Tertiary Industry		
2001	9 549. 9	646. 3	4 199. 0	4 704. 6	8. 7	
2002	11 305. 9	730. 0	4 920. 2	5 655. 7	9. 4	19. 8
2003	11 983. 7	766. 1	5 422. 3	5 795. 3	8. 8	4. 1
2004	14 697. 7	851. 0	6 726. 6	7 120. 1	9. 2	16. 8
2005	17 698. 5	1 008. 9	8 123. 6	8 566. 0	9. 6	16. 3
2006	21 220. 3	1 142. 9	9 802. 5	10 274. 9	10. 1	16. 3

2-2 全国海洋生产总值构成

Composition of National Gross Ocean Product

年 份 Year	第一产业 Primary Industry （%）	第二产业 Secondary Industry （%）	第三产业 Tertiary Industry （%）
2001	6. 8	44. 0	49. 2
2002	6. 5	43. 5	50. 0
2003	6. 4	45. 2	48. 4
2004	5. 8	45. 8	48. 4
2005	5. 7	45. 9	48. 4
2006	5. 4	46. 2	48. 4

2-3 海洋及相关产业增加值
Added Values of Marine and Related Industries

单位：亿元 Unit:100 million yuan

年 份 Year	海洋生产总值 Gross Ocean Product	海洋产业 Marine Industry	主要海洋产业 Major Marine Industry	海洋科研教育管理服务业 Industries of Marine Scientific Research, Education, Management and Service	海洋相关产业 Ocean-related Industries
2001	9 549.9	5 714.1	3 856.6	1 857.5	3 835.8
2002	11 305.9	6 764.0	4 696.8	2 067.2	4 541.9
2003	11 983.7	7 109.8	4 754.4	2 355.4	4 873.9
2004	14 697.7	8 676.6	5 827.7	2 849.0	6 021.1
2005	17 698.5	10 498.6	7 188.0	3 310.5	7 199.9
2006	21 220.3	12 548.2	8 792.5	3 755.8	8 672.1

2-4 海洋及相关产业增加值构成
Composition of the Added Value of Marine and Related Industries

年 份 Year	海洋生产总值（%） Gross Ocean Product（%）	海洋产业 Marine Industry	主要海洋产业 Major Marine Industry	海洋科研教育管理服务业 Industries of Marine Scientific Research, Education, Management and Service	海洋相关产业 Ocean-related Industries
2001	100.0	59.8	40.4	19.5	40.2
2002	100.0	59.8	41.5	18.3	40.2
2003	100.0	59.3	39.7	19.7	40.7
2004	100.0	59.0	39.7	19.4	41.0
2005	100.0	59.3	40.6	18.7	40.7
2006	100.0	59.1	41.4	17.7	40.9

2-5 全国主要海洋产业增加值（2006年）
Gross Output Value and Added Value of Major Marine Industries, 2006

主要海洋产业 Major Marine Industry	增加值 (亿元) Added Value (100 million yuan)	增加值比上年增长（%） (按可比价计算) Percentage of Added Value of Increase Over Last Year (%) (at comparable price)
合 计 **Total**	**8 792.5**	**17.6**
海洋渔业 Marine Fishery Industry	1 708. 1	6. 6
海洋石油和天然气 Offshore Oil and Natural Gas	668. 9	3. 8
海洋矿业 Marine Mining Industry	6. 3	-38. 2
海洋盐业 Sea Salt Industry	40. 9	2. 3
海洋化工 Marine Chemical	184. 6	20. 0
海洋生物医药 Marine Biological Pharmaceatical	28. 2	0. 1
海洋电力 Marine Electric Power	4. 4	22. 3
海水利用 Marine Seawater Utilization	3. 4	7. 7
海洋船舶工业 Marine Shipbuilding Industry	380. 6	38. 8
海洋工程建筑 Marine Engineering Architecture	305. 5	17. 2
海洋交通运输 Maritime Transportation	2 842. 1	16. 0
滨海旅游 Coastal Tourism	2 619. 6	29. 0

2-6 海洋渔业增加值
Added Value of Marine Fishery

单位：亿元 Unit:100 million yuan

年 份 Year	增加值 Added Value
2001	966.0
2002	1 091.2
2003	1 145.0
2004	1 271.2
2005	1 507.6
2006	1 708.1

2-7 海洋石油和天然气业增加值
Added Value of Offshore Oil and Gas Industry

单位：亿元 Unit:100 million yuan

年 份 Year	增加值 Added Value
2001	176.8
2002	181.8
2003	257.0
2004	345.1
2005	528.2
2006	668.9

2-8 海洋矿业增加值
Added Value of Marine Mining Industry

单位：亿元 Unit:100 million yuan

年 份 Year	增加值 Added Value
2001	1.0
2002	1.9
2003	3.1
2004	7.9
2005	8.3
2006	6.3

2-9 海洋盐业增加值
Added Value of Marine Salt Industry

单位：亿元 Unit:100 million yuan

年 份 Year	增加值 Added Value
2001	32.6
2002	34.2
2003	28.4
2004	39.0
2005	39.1
2006	40.9

2-10 海洋化工业增加值
Added Value of Marine Chemical Industry

单位：亿元 Unit:100 million yuan

年　份 Year	增加值 Added Value
2001	64. 7
2002	77. 1
2003	96. 3
2004	151. 5
2005	153. 3
2006	184. 6

2-11 海洋生物医药业增加值
Added Value of Marine Biopharmaceutics Industry

单位：亿元 Unit:100 million yuan

年　份 Year	增加值 Added Value
2001	5. 7
2002	13. 2
2003	16. 5
2004	19. 0
2005	28. 6
2006	28. 2

2-12 海洋电力业增加值
Added Value of Marine Electric Power Industry

单位：亿元 Unit:100 million yuan

年 份 Year	增加值 Added Value
2001	1.8
2002	2.2
2003	2.8
2004	3.1
2005	3.5
2006	4.4

2-13 海水利用业增加值
Added Value of Seawater Utilization Industry

单位：亿元 Unit:100 million yuan

年 份 Year	增加值 Added Value
2001	1.1
2002	1.3
2003	1.7
2004	2.4
2005	3.0
2006	3.4

2-14 海洋船舶工业增加值
Added Value of Marine Shipbuilding Industry

单位：亿元 Unit:100 million yuan

年 份 Year	增加值 Added Value
2001	109.3
2002	117.4
2003	152.8
2004	204.1
2005	275.5
2006	380.6

2-15 海洋工程建筑业增加值
Added Value of Marine Engineering Architecture

单位：亿元 Unit:100 million yuan

年 份 Year	增加值 Added Value
2001	109.2
2002	145.4
2003	192.6
2004	231.8
2005	257.2
2006	305.5

2-16 海洋交通运输业增加值

Added Value of Marine Transportation Industry

单位：亿元 Unit:100 million yuan

年　份 Year	增加值 Added Value
2001	1 316.4
2002	1 507.4
2003	1 752.5
2004	2 030.7
2005	2 373.3
2006	2 842.1

2-17 滨海旅游业增加值

Added Value of Coastal Tourism

单位：亿元 Unit:100 million yuan

年　份 Year	增加值 Added Value
2001	1 072.0
2002	1 523.7
2003	1 105.8
2004	1 522.0
2005	2 010.6
2006	2 619.6

2-18 沿海地区海洋生产总值（2006年）

Gross Ocean Product in the Coastal Regions, 2006

地区 Region	海洋生产总值（亿元） Gross Ocean Product (100 million yuan)	第一产业 Primary Industry	第二产业 Secondary Industry	第三产业 Tertiary Industry	海洋生产总值占沿海地区生产总值比重（%） Proportion of the Gross Ocean Product in the Gross Regional Product
合计 Total	**21 220.3**	**1 142.9**	**9 802.5**	**10 274.9**	**15.7**
天津 Tianjin	1 369.0	3.5	900.9	464.6	31.4
河北 Hebei	1 092.1	24.8	554.0	513.4	9.4
辽宁 Liaoning	1 478.9	146.4	791.2	541.3	16.0
上海 Shanghai	3 988.2	3.5	1 924.2	2 060.6	38.5
江苏 Jiangsu	1 287.0	65.4	547.2	674.5	5.9
浙江 Zhejiang	1 856.5	137.8	736.1	982.6	11.8
福建 Fujian	1 743.1	169.2	701.3	872.6	22.9
山东 Shandong	3 679.3	306.9	1 786.4	1 585.9	16.7
广东 Guangdong	4 113.9	182.8	1 640.5	2 290.6	15.7
广西 Guangxi	300.7	45.7	129.7	125.4	6.2
海南 Hainan	311.6	57.0	91.0	163.5	29.6

2-19 沿海地区海洋生产总值构成（2006年）

Composition of Gross Ocean Product in the Coastal Regions, 2006

地区 Region	海洋生产总值（%） Gross Ocean Product（%）	第一产业 Primary Industry	第二产业 Secondary Industry	第三产业 Tertiary Industry
合计 Total	**100.0**	**5.4**	**46.2**	**48.4**
天津 Tianjin	100.0	0.3	65.8	33.9
河北 Hebei	100.0	2.3	50.7	47.0
辽宁 Liaoning	100.0	9.9	53.5	36.6
上海 Shanghai	100.0	0.1	48.2	51.7
江苏 Jiangsu	100.0	5.1	42.5	52.4
浙江 Zhejiang	100.0	7.4	39.7	52.9
福建 Fujian	100.0	9.7	40.2	50.1
山东 Shandong	100.0	8.3	48.6	43.1
广东 Guangdong	100.0	4.4	39.9	55.7
广西 Guangxi	100.0	15.2	43.1	41.7
海南 Hainan	100.0	18.3	29.2	52.5

2-20 沿海地区海洋及相关产业增加值（2006年）
Added Values of Marine and Related Industries by Coastal Regions, 2006

单位：亿元 Unit:100 million yuan

地 区 Region	海洋生产总值 Gross Ocean Product				
		海洋产业 Marine Industry	主要海洋产业 Major Marine Industry	海洋科研教育管理服务业 Industries of Marine Scientific Research, Education, Management and Service	海洋相关产业 Ocean-related Industries
合 计 Total	**21 220.3**	**12 548.2**	**8 792.5**	**3 755.8**	**8 672.1**
天 津 Tianjin	1 369.0	753.5	684.7	68.7	615.5
河 北 Hebei	1 092.1	608.0	561.4	46.6	484.2
辽 宁 Liaoning	1 478.9	875.7	695.2	180.5	603.3
上 海 Shanghai	3 988.2	1 996.1	1 305.0	691.2	1 992.1
江 苏 Jiangsu	1 287.0	820.6	547.7	272.9	466.4
浙 江 Zhejiang	1 856.5	1 176.0	763.8	412.2	680.5
福 建 Fujian	1 743.1	1 037.4	735.7	301.8	705.7
山 东 Shandong	3 679.3	2 149.4	1 625.6	523.8	1 529.9
广 东 Guangdong	4 113.9	2 725.7	1 573.0	1 152.7	1 388.2
广 西 Guangxi	300.7	184.9	142.5	42.4	115.8
海 南 Hainan	311.6	221.0	158.0	62.9	90.6

2-21 沿海地区海洋及相关产业增加值构成（2006年）

Composition of the Added Value of Marine and Related Industries by Coastal Regions, 2006

地区 Region	海洋生产总值（%） Gross Ocean Product（%）	海洋产业 Marine Industry	主要海洋产业 Major Marine Industry	海洋科研教育管理服务业 Industries of Marine Scientific Research, Education, Management and Service	海洋相关产业 Ocean-related Industries
合计 Total	**100.0**	**59.1**	**41.4**	**17.7**	**40.9**
天津 Tianjin	100.0	55.0	50.0	5.0	45.0
河北 Hebei	100.0	55.7	51.4	4.3	44.3
辽宁 Liaoning	100.0	59.2	47.0	12.2	40.8
上海 Shanghai	100.0	50.0	32.7	17.3	49.9
江苏 Jiangsu	100.0	63.8	42.6	21.2	36.2
浙江 Zhejiang	100.0	63.3	41.1	22.2	36.7
福建 Fujian	100.0	59.5	42.2	17.3	40.5
山东 Shandong	100.0	58.4	44.2	14.2	41.6
广东 Guangdong	100.0	66.2	38.2	28.0	33.7
广西 Guangxi	100.0	61.5	47.4	14.1	38.5
海南 Hainan	100.0	70.9	50.7	20.2	29.1

主要统计指标解释

1.海洋经济 是开发、利用和保护海洋的各类产业活动,以及与之相关联活动的总和。

2.海洋生产总值 是海洋经济生产总值的简称,指按市场价格计算的沿海地区常住单位在一定时期内海洋经济活动的最终成果,是海洋产业和海洋相关产业增加值之和。

3.海洋产业 是开发、利用和保护海洋所进行的生产和服务活动,包括海洋渔业、海洋油气业、海洋矿业、海洋盐业、海洋化工业、海洋生物医药业、海洋电力业、海水利用业、海洋船舶工业、海洋工程建筑业、海洋交通运输业、滨海旅游业等主要海洋产业,以及海洋科研教育管理服务业。

4.海洋科研教育管理服务业 是开发、利用和保护海洋过程中所进行的科研、教育、管理及服务等活动,包括海洋信息服务业、海洋环境监测预报服务、海洋保险与社会保障业、海洋科学研究、海洋技术服务业、海洋地质勘查业、海洋环境保护业、海洋教育、海洋管理、海洋社会团体与国际组织等。

5.海洋相关产业 是指以各种投入产出为联系纽带,与主要海洋产业构成技术经济联系的上下游产业,涉及海洋农林业、海洋设备制造业、涉海产品及材料制造业、涉海建筑与安装业、海洋批发与零售业、涉海服务业等。

6.海洋三次产业 我国的海洋三次产业划分是:

海洋第一产业:是指海洋渔业中的海洋水产品、海洋渔业服务业,以及海洋相关产业中属于第一产业范畴的部门。

海洋第二产业:是指海洋渔业中海洋水产品加工、海洋油气业、海洋矿业、海洋盐业、海洋化工业、海洋生物医药业、海洋电力业、海水利用业、海洋船舶工业、海洋工程建筑业,以及海洋相关产业中属于第二产业范畴的部门。

海洋第三产业:是指除海洋第一、二产业以外的其他行业。第三产业包括:海洋交通运输业、滨海旅游业、海洋科研教育管理服务业,以及海洋相关产业中属于第三产业范畴的部门。

7.海洋渔业 包括海水养殖、海洋捕捞、海洋渔业服务业和海洋水产品加工等活动。

8.海洋油气业 是指在海洋中勘探、开采、输送、加工原油和天然气的生产活动。

9.海滨砂矿业 包括海滨砂矿、海滨土砂石、海底地热、煤矿开采和深海采矿等采选活动。

10.海洋盐业 是指利用海水生产以氯化钠为主要成分的盐产品的活动,包括采盐和盐加工。

11.海洋化工业 包括海盐化工、海水化工、海藻化工及海洋石油化工的化工产品生产活动。

12.海洋生物医药业 是指以海洋生物为原料或提取有效成分,进行海洋药品与海洋保健品的生产加工及制造活动。

13.海洋电力业 是指在沿海地区利用海洋能、海洋风能进行的电力生产活动。不包括沿海地区的火力发电和核力发电。

14.海水利用业 是指对海水的直接利用和海水淡化活动,包括利用海水进行淡水生产和将海水应用于工业冷却用水和城市生活用水、消防用水等活动,不包括海水化学资源综合利用活动。

15.海洋船舶工业 是指以金属或非金属为主要材料，制造海洋船舶、海上固定及浮动装置的活动，以及对海洋船舶的修理及拆卸活动。

16.海洋工程建筑业 是指在海上、海底和海岸所进行的用于海洋生产、交通、娱乐、防护等用途的建筑工程施工及其准备活动；包括海港建筑、滨海电站建筑、海岸堤坝建筑、海洋隧道桥梁建筑、海上油气田陆地终端及处理设施建造、海底线路管道和设备安装，不包括各部门、各地区的房屋建筑及房屋装修工程。

17. 海洋交通运输业 是指以船舶为主要工具从事海洋运输以及为海洋运输提供服务的活动，包括远洋旅客运输、沿海旅客运输、远洋货物运输、沿海货物运输、水上运输辅助活动、管道运输业、装卸搬运及其他运输服务活动。

18.滨海旅游业 是指以海岸带、海岛及海洋各种自然景观、人文景观为依托的旅游经营、服务活动，主要包括:海洋观光游览、休闲娱乐、度假住宿、体育运动等活动。

Explanatory Notes on Main Statistical Indicators

1.Marine Economy is the summation of various types of industrial activities for developing, utilizing and protecting the ocean as well as the activities associated therewith.

2.Gross Ocean Product is the short form of the gross output value of ocean economy, referring to the final result of marine economic activities of the permanent units in the coastal region within a given period calculated at the market price, and the sum total of the added values of the marine industries and the ocean-related industries.

3.Marine industry refers to the production as service activities for developing, utilizing and protecting the ocean, including major marine industries such as offshore oil and gas industry, marine mining industry, marine salt industry, marine chemical industry, marine biopharmaceutics industry, marine electric power industry, seawater utilization industry, marine shipbuilding industry, marine engineering construction industry, marine communications industry, coastal tourism etc. as well as marine scientific research, education, management and service.

4.Marine Scientific Research, Education, Management and Service refers to the activities of scientific research, education, management and service carried out in the process of developing, utilizing and protecting the ocean, including marine information service industry, marine environment monitoring and forecasting service, marine insurance and social security industry, marine scientific research, marine technological service industry, ocean geological prospecting industry, marine environmental protection industry, marine education, marine management,marine social organization and international organizations,etc.

5.Ocean–Related Industry refers to the lower and upper reaches enterprises that form a technical and economic link with the major marine industries, with various inputs and outputs as ties, involving marine agriculture and forestry, marine equipment manufacturing, ocean-related building and installation industry, marine wholesale and retail industry, ocean-related service industry,etc.

6.Marine Three Industries Chinese marine three industries are divided as follows:

Marine primary Industry: refers to the marine aquatic products, marine fishery service industry in the marine fishery as well as the sectors belonging to the primary industry category in the ocean-related industries.

Marine secondary industry: refers to the marine aquatic products processing industry in the marine fishery,offshore oil as gas industry, marine mining industry, marine salt industry, marine chemical industry, marine biopharmaceutics industry, marine electric power industry, seawater utilization industry, marine shipbuilding industry, marine engineering construction industry, as well as the sectors belonging to the category of secondary industry in the ocean-related industries.

Marine tertiary industry: refers to the industries other than the marine primary and secondary industries, including marine transportation and communication industry, coastal tourism, marine scientific research, education, management and service industry as well as the sectors belonging to the category of tertiary industry in the ocean-related industries.

7.Marine Fishery includes mariculture, marine fishing, marine fishery service industry and marine aquatic products processing,etc.

8.Offshore Oil and Gas Industry refers to the production activities of exploring, exploiting, transporting and processing crude oil and natural gas in the ocean.

9.Beach Placer Mining Industry includes the activities of extracting and dressing beach placers, beach soil and sand, submarine geothermal heat, coal mining and deep-sea mining,etc.

10.Marine Salt Industry refers to the activity of producing salt products with the sodium chloride as the main component by utilizing seawater, including salt extracting and salt processing.

11.Marine Chemical Industry includes the productive activities of chemical products of sea salt, seawater, sea algal and marine petroleum chemical industries.

12.Marine Biopharmaceutics Industry refers to the production, processing and manufacturing activities of marine medicines and marine health care products by using marine organisms as raw materials or extracting useful components therefrom.

13.Marine Electric Power Industry refers to the activities of generating electric power in the coastal region by making use of ocean energies and ocean wind energy. It does not include the thermal and nuclear power generation in the coastal area.

14.Seawater Utilization Industry refers to the activities of the direct use of sea water and the seawater desalination, including those of carrying out the production of desalination and applying the seawater as water for industrial cooling, urban domestic water, water for fire fighting etc., but not the activity of the multipurpose use of seawater chemical resources.

15.Shipbuilding Industry refers to the activity of building ocean vessels, offshore fixed and floating equipment with metals or non-metals as main materials as well as repairing and dismantling ocean vessels.

16.Marine Engineering Building Industry refers to the architectural projects construction and its preparations in the sea, at the sea bottom and seacoast for such uses as marine production, transportation, recreation, protection, etc., including constructions of seaports, coastal power stations, coastal dykes, marine tunnels and bridges, land terminals of offshore oil and gas fields as well as building of processing facilities, and installation of submarine pipelines and equipment, but not the projects of house building and renovation.

17.Marine Transportation and Communications Industry refers to the activities of carrying out and serving the sea transportations with vessels as main vehicles, including ocean-going passagers transportation, coastal passagers transportation, occan-going cargo transportation, coastal cargo transportation, auxiliary activities of water transportation, pipeline transportation, loading, unloading and transport as well as other transportation service activities.

18.Coastal Tourism refers to the tourist business and service activities with the backing of coastal zone, sea islands as well as a variety of natural and human landscapes of the ocean, mainly including marine sightseeing, living a life of leisure and receation, going on vocation and getting accommodation, sports, etc.

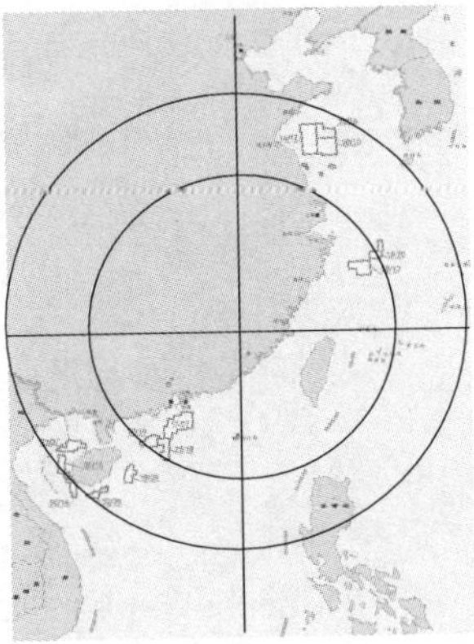

3

主要海洋产业活动

Major Marine Industrial Activities

3-1 全国海洋捕捞养殖产量
National Marine Fishery Production

项　目　Item	2004	2005	2006
海水水产品产量（吨）Total Seawater Aquatic Products (ton)	**27 677 907**	**28 380 831**	**28 876 758**
海洋捕捞产量 Marine Catches	**14 510 858**	**14 532 984**	**14 420 359**
按品种分　By Species			
鱼类　Fish	9 590 111	9 860 826	9 658 444
甲壳类　Crustacea	2 401 850	2 409 410	2 470 957
贝类　Shellfish	846 868	929 319	903 832
藻类　Algae	37 671	30 367	31 822
头足类　Cephalopoda	1 141 474	1 029 620	961 659
其他　Others	492 884	275 942	393 645
海水养殖产量 Mariculture Production	**13 167 049**	**13 847 847**	**14 456 399**
按品种分　By Species			
鱼类　Fish	582 566	658 928	715 275
甲壳类　Crustacea	722 172	828 511	937 563
贝类　Shellfish	10 247 151	10 674 762	11 135 933
藻类　Algae	1 467 545	1 511 285	1 502 804
其他　Others	147 615	174 361	164 824

3-2 沿海地区海洋捕捞养殖产量（2006年）

Marine Fishing and Mariculture Production by Coastal Regions, 2006

单位：吨 Unit:ton

地区 Region	海洋捕捞产量 Marine Catches	#远洋捕捞产量 Deep-Sea Fishing Production	海水养殖产量 Mariculture Production
全国总计 National Total	**14 420 359**	**1 090 663**	**14 456 399**
天津 Tianjin	32 827	10 820	16 457
河北 Hebei	313 734	2 900	304 765
辽宁 Liaoning	1 481 696	180 065	2 223 919
上海 Shanghai	193 987	162 951	417
江苏 Jiangsu	584 386	13 956	606 369
浙江 Zhejiang	3 119 084	158 398	886 147
福建 Fujian	2 169 678	108 963	3 118 800
山东 Shandong	2 574 098	81 088	3 735 008
广东 Guangdong	1 685 640	111 170	2 419 536
广西 Guangxi	839 773	0	929 735
海南 Hainan	1 165 104	0	215 246

3-3 沿海地区海洋原油产量
Output of Offshore Crude Oil by Coastal Regions

单位：万吨　Unit:10 000 tons

地　区 Region	2004	2005	2006
合　计　Total	**2 842.21**	**3 174.66**	**3 239.91**
天　津　Tianjin	1 094.93	1 311.29	1 479.16
河　北　Hebei		126.03	155.79
辽　宁　Liaoning	19.99	18.99	19.03
上　海　Shanghai	31.90	25.58	21.69
山　东　Shandong	213.50	212.00	216.00
广　东　Guangdong	1 481.89	1 480.77	1 348.24

3-4 沿海地区海洋天然气产量
Output of Offshore Natural Gas by Coastal Regions

单位：万立方米　Unit:10 000 m^3

地　区 Region	2004	2005	2006
合　计　Total	**613 416**	**626 921**	**748 618**
天　津　Tianjin	87 188	81 869	104 958
辽　宁　Liaoning	9 911	8 200	8 561
上　海　Shanghai	62 076	66 336	72 608
山　东　Shandong	18 833	16 353	13 850
广　东　Guangdong	435 408	454 163	548 641

3-5 海洋原油出口量及创汇额

Export Volume and Foreign-Exchange Earnings of Offshore Crude Oil by Coastal Regions

单位:万吨, 万美元

Unit: 10 000 tons, 10 000 US$

地 区 Region	2004		2005		2006	
	出口量 Export Volume	创汇额 Foreign-Exchange Earnings	出口量 Export Volume	创汇额 Foreign-Exchange Earnings	出口量 Export Volume	创汇额 Foreign-Exchange Earnings
合 计 Total	**457.68**	**116 174**	**679.37**	**235 952**	**567.62**	**234 689**
天 津 Tianjin	83.74	16 837	207.51	60 309	230.50	75 629
广 东 Guangdong	373.94	99 337	471.86	175 643	337.12	159 060

3-6 海洋原油产量、出口量占全国原油产量、出口量比重

Proportion of Offshore Crude Oil Production and Export Volume in the National Total

年 份 Year	海洋原油产量占全国原油产量比重(%) Proportion of Offshore Crude Oil Production in the National Total(%)	海洋原油出口量占全国原油出口量比重(%) Proportion of Offshore Crude Oil Export Volume in the National Total(%)
2001	13.07	46.26
2002	14.40	54.95
2003	15.01	60.77
2004	16.16	83.37
2005	17.51	84.18
2006	17.54	89.53

3-7 沿海地区海滨砂矿产量（2006年）
Output of Beach Placer by Coastal Regions, 2006

单位：吨　Unit: ton

地　区 Region	产　量 Output
合　计 Total	**103 552 172**
浙　江 Zhejiang	41 666 223
福　建 Fujian	2 104 900
山　东 Shandong	300 000
广　东 Guangdong	150 000
广　西 Guangxi	382 070
海　南 Hainan	58 948 979

3-8 沿海地区海盐产量
Productive Activities of Salt-Making Industry by Coastal Regions

单位：万吨　　Unit: 10 000 tons

地　区 Region	海盐产量 Output of Sea Salt		
	2004	2005	2006
合　计 Total	**2 307.28**	**2 828.55**	**3 100.89**
天　津 Tianjin	238.50	230.60	236.10
河　北 Hebei	394.42	461.92	395.07
辽　宁 Liaoning	204.00	212.13	226.7*
江　苏 Jiangsu	187.65	186.90	168.63
浙　江 Zhejiang	52.15	38.68	28.96
福　建 Fujian	57.33	43.87	49.59
山　东 Shandong	1 132.24	1 610.04	1 950.96*
广　东 Guangdong	23.54	20.19*	18.59*
广　西 Guangxi	8.91	8.02	8.04
海　南 Hainan	8.54	16.20	18.28

注：*为中国盐业总公司数据。

Note: * is the data for the China Salt Industry Corporation.

3-9 沿海地区海洋化工产品产量
Output of Marine Chemical Products by Coastal Regions

单位：吨 Unit:ton

地 区 Region	产品产量* Output		
	2004	2005	2006
合 计 Total	**6 019 480**	**4 722 759**	**10 001 784**
天 津 Tianjin	2 446 221	1 659 203	1 864 755
河 北 Hebei	304 481	244 082	1 759 694
辽 宁 Liaoning	1 256 907	1 267 300	
江 苏 Jiangsu	1 608 101	168 588	262 921
浙 江 Zhejiang	165 588	201 124	
山 东 Shandong	234 682	1 182 462	6 114 414
广 东 Guangdong	3 500		

注：本部分为沿海地区部分海洋化工企业产品汇总数据。
Note: This part is the data collected from the products of part of the chemical enterprises in the coastal region.

3-10 沿海地区海洋生物医药产品产量（2006年）
Production of Marine Biopharmaceutics Industry by Coastal Regions, 2006

产品名称 Name	计量单位 Unit	产品产量 Output
甲壳素Crustaceoxin	吨 ton	3 313
海参肽营养胶囊Sea Cucumber Peptide	箱 box	22 890
医药输液Drip Liquid	万瓶 10 000 bottles	9 631
海藻素Pheophytin	吨 ton	36 771
多烯酸乙酯胶囊Ethyl Polyenoic Acid Capsule	万粒 10 000 pellets	2 268
鲨鱼肝油胶丸Shark Cod-liver oil pill	万粒 10 000 pellets	2 837
金枪鱼钙胶丸Tuna Calcium pill	万粒 10 000 pellets	1 725
卵磷脂 Lecithin	万粒 10 000 pellets	1 769
鲨鱼软骨胶囊Shark Cartilage Capsule	万粒 10 00 pellets	65
海默止血片Haimo Haemostatic Pill	万盒 10 000 cases	7
角鲨烯胶囊Houndfish Alkene Capsule	万粒 10 000 pellets	1 013
脑元神胶囊Naoyuanshen Capsule	万粒 10 000 pellets	322
珍珠明目滴眼液Zhenzhu Eye-Brightening Drops	万瓶 10 000 bottles	166
氨糖美辛片Meixin Amino Sugar	万粒 10 000 pellets	1 960
智力宝、心脑灵Zhilibao,Xinnaoling	箱 box	20 680
藻日康Zaorikang	盒 case	79 000
氨基酸盐Amino-Acid Salr	万吨 10 000 tons	1
海威口服液Haiwei Oral Liguid	吨 ton	260
螺旋藻片、胶囊Spiraling Pill Capsule	瓶 bottles	720 000
牡蛎钙颗粒Oyster-Calcium Peller	万袋 10 000 bags	16
牡蛎钙片Oyster-Calcium Pill	万片 10 000 pills	147
银耳鱼肝油乳Tremella Cod-Liver Oil Milk	万瓶 10 000 bottles	50
三维鱼肝油乳Sanwei Cod-Liver Oil Milk	万瓶 10 000 bottles	189
鱼肝油乳 Cod-Liver Oil Milk	万瓶 10 000 bottles	256
维生素AD胶囊Vitamin AD Capsule	万瓶 10 000 bottles	109 378
维生素AD滴剂Vitamin AD Drops	万瓶 10 000 bottles	295
鲎试剂King Crab Tablet	万支 10 000 bottles	41
海珠喘息定片Haizhu Methoxyphenamine Pill	万片 10 000 pills	13 500
珍珠粉末Pearl Powder	万瓶 10 000 bottles	510
藻酸双脂钠Alginic Acid Diadipose Sodium	万粒 10 000 pellets	76
麝珠明目滴眼液Shezhu Eye-Brightening Drops	万支 10 000 bottles	104
琼脂Agar	吨 ton	294

注：此表数据为沿海地区部分海洋生物医药企业产品汇总数据。

Note: The data in this table are those collected from part of the marine biopharmaceutical enterprises in the coastal region.

3-11 沿海地区海洋修造船完工量（2006年）
Production of the Marine Shipbuilding Industry by Coastal Regions, 2006

部门和地区 Sector and Region	修船完工量（艘） Ships Repaired (number)	造船完工量 Ships Built 艘 number	 万综合吨 Comprehensive Tonnages (10 000 tons)
总 计 Total	**27 262**	**2 358**	**1 350.90**
其 中: Including:			
中船工业集团公司 CSSC	323	94	575.24
中船重工集团公司 CSIC	351	65	264.66
按地区分: By Regions:			
天 津 Tianjin	97	10	9.00
河 北 Hebei	128	2	6.00
辽 宁 Liaoning	551	54	265.00
上 海 Shanghai	1 015	69	516.42
江 苏 Jiangsu	2 756	141	202.87
浙 江 Zhejiang	12 436	951	214.01
福 建 Fujian	1 716	195	46.05
山 东 Shandong	8 393	906	32.39
广 东 Guangdong	128	30	59.16
广 西 Guangxi			
海 南 Hainan	42	0	0.00

3-12 沿海地区海洋货物运输量和周转量（2006年）
Maritime Freight Volume and Turnover Volume by Coastal Regions, 2006

单位：万吨，亿吨公里

Unit:10 000 tons,100 million ton-km

地区 Region	货运量 Freight Volume	沿海 Coastal	远洋 Oceangoing	货物周转量 Cargo Turnover Volume	沿海 Coastal	远洋 Oceangoing
合计 Total	**132 613**	**78 200**	**54 413**	**52 460.44**	**9 883.13**	**42 577.30**
天津 Tianjin	13 239	2 070	11 169	11 711.60	430.74	11 280.85
河北 Hebei	2 778	1 112	1 666	2 051.41	316.41	1 735.00
辽宁 Liaoning	7 201	3 525	3 676	2 361.31	357.22	2 004.10
上海 Shanghai	33 810	22 062	11 748	13 608.95	2 904.72	10 704.23
江苏 Jiangsu	7 083	4 578	2 505	1 931.16	390.28	1 540.88
浙江 Zhejiang	23 043	22 090	953	3 312.46	2 565.48	746.98
福建 Fujian	8 235	7 336	899	1 421.65	1 090.25	331.40
山东 Shandong	8 431	2 421	6 010	4 083.61	238.74	3 844.87
广东 Guangdong	13 976	6 735	7 241	2 832.66	1 065.34	1 767.32
广西 Guangxi	581	350	231	47.48	33.99	13.50
海南 Hainan	5 359	4 640	719	568.85	380.68	188.17
其他 Others	8 877	1 281	7 596	8 529.30	109.28	8 420.02

3-13 沿海地区海洋国际标准集装箱运量

Seaborne Volume of International Standardized Containers Traffic by Coastal Regions

单位：万 TEU, 万吨

Unit：10 000 TEU,10 000 tons

地区 Region	2004		2005		2006	
	箱数 Containers	重量 Weight	箱数 Containers	重量 Weight	箱数 Containers	重量 Weight
合计 Total	**1 605.3**	**15 938.2**	**1 940.0**	**21 991.6**	**2 329.7**	**25 774.7**
天津 Tianjin			28.6	414.0	21.5	297.0
河北 Hebei					0.7	1.5
辽宁 Liaoning	13.5	110.2	28.5	246.3	24.4	215.6
上海 Shanghai	933.7	9 946.9	1 130.7	13 739.1	1 307.7	15 756.4
江苏 Jiangsu	32.5	215.7	54.2	393.4	65.8	553.4
浙江 Zhejiang	20.7	266.3	18.2	255.1	31.3	463.3
福建 Fujian	54.7	537.9	60.0	571.9	88.4	881.5
山东 Shandong	85.7	978.8	103.5	1 364.8	155.6	1 510.2
广东 Guangdong	387.1	2 915.3	422.8	3 777.2	451.0	3 837.6
广西 Guangxi	14.2	166.9	9.5	78.4	12.1	113.6
海南 Hainan	15.9	260.1	16.5	264.2	57.5	884.2
其他 Others	47.4	540.0	67.5	887.4	113.7	1 260.2

注：本表数据包含内河集装箱数。

Note: The data in this table include the freight volume of containers in the inland rivers.

3-14 沿海地区海洋旅客运输量和周转量（2006年）

Maritime Passenger Traffic and Turnover Volume by Coastal Regions, 2006

单位：万人，亿人公里
Unit:10 000 persons,100 million passenger-km

地 区 Region	客运量 Passenger Traffic	沿 海 Coastal	远 洋 Oceangoing	旅客周转量 Passenger Turnover Volume	沿 海 Coastal	远 洋 Oceangoing
合 计 Total	**9 860**	**9 080**	**780**	**42.09**	**30.88**	**11.21**
天 津 Tianjin	4	-	4	0. 39		0. 39
辽 宁 Liaoning	714	704	10	9. 16	8. 69	0. 47
上 海 Shanghai	1 263	1 261	2	4. 59	4. 33	0. 26
浙 江 Zhejiang	2 406	2 406	-	5. 96	5. 96	-
福 建 Fujian	952	918	34	1. 30	1. 20	0. 10
山 东 Shandong	1 417	1 373	44	6. 63	4. 53	2. 10
广 东 Guangdong	1 639	953	686	11. 41	3. 52	7. 89
广 西 Guangxi	61	61	-	0. 56	0. 56	-
海 南 Hainan	1 404	1 404	-	2. 10	2. 10	-

3-15 沿海地区主要港口客货吞吐量（2006年）

Passengers Leaving and Arriving and Cargo Handled at Principal Seaports by Coastal Regions, 2006

单位：万吨，万人次

Unit：10 000 tons，10 000 persons

地　区 Region	货物吞吐量 Cargo Handled	#外　贸 Foreign Trade	旅客吞吐量 Passenger Leaving & Arriving	#离　港 Leaving
合　计　Total	**353 031**	**147 548**	**7 479**	**3 763**
天　津　Tianjin	25 760	13 966	34	15
河　北　Hebei	33 805	8 212	5	2
辽　宁　Liaoning	35 658	9 982	630	318
上　海　Shanghai	47 040	21 268	1 324	654
江　苏　Jiangsu	7 683	4 526	6	3
浙　江　Zhejiang	51 281	19 244	2 117	1 060
福　建　Fujian	23 688	7 912	68	34
山　东　Shandong	47 006	28 720	862	434
广　东　Guangdong	70 717	29 575	1 865	942
广　西　Guangxi	4 949	3 481	74	38
海　南　Hainan	5 444	662	492	262

3-16 沿海地区主要港口国际标准集装箱吞吐量
International Standardized Containers Handled at Principal Seaports by Coastal Regions

单位: 万TEU,万吨
Unit: 10 000TEU,10 000 tons

地 区 Region	2004		2005		2006	
	箱 数 Containers	重 量 Weight	箱 数 Containers	重 量 Weight	箱 数 Containers	重 量 Weight
合 计 Total	**5 656.7**	**50 966.7**	**6 988.8**	**63 352.0**	**8 563.4**	**80 077.0**
天 津 Tianjin	381. 6	4 314. 6	480. 1	5 352. 8	595. 0	6 486. 0
河 北 Hebei	8. 8	101. 2	14. 0	166. 5	29. 5	383. 0
辽 宁 Liaoning	298. 9	2 908. 5	377. 9	3 956. 5	468. 4	5 846. 0
上 海 Shanghai	1 455. 4	13 293. 7	1 808. 5	16 250. 4	2 171. 9	19 595. 0
江 苏 Jiangsu	50. 8	505. 3	101. 6	1 074. 9	131. 2	1 307. 0
浙 江 Zhejiang	432. 7	3 724. 4	555. 5	4 818. 4	751. 9	6 760. 0
福 建 Fujian	425. 7	4 192. 4	492. 5	4 895. 3	587. 3	6 205. 0
山 东 Shandong	576. 8	5 556. 4	738. 6	7 065. 6	936. 8	9 032. 0
广 东 Guangdong	1 990. 4	15 890. 4	2 378. 1	19 208. 7	2 837. 6	23 733. 0
广 西 Guangxi	11. 8	150. 4	15. 4	191. 6	20. 9	279. 0
海 南 Hainan	23. 7	329. 5	26. 6	371. 4	32. 8	453. 0

3-17 沿海城市国内旅游人数
Domestic Tourists by Coastal Cities

单位：万人 Unit:10 000 persons

城 市	City	2004	2005	2006
合 计	**Total**	**36 334**	**38 574**	**50 717**
天 津	**Tianjin**	**3 751**	**4 484**	**5 013**
河 北	**Hebei**	**1 040**	**1 901**	**2 098**
唐 山	Tangshan	338	510	565
秦皇岛	Qinhuangdao	592	1 212	1 302
沧 州	Cangzhou	110	179	231
辽 宁	**Liaoning**	**2 558**	**3 343**	**4 061**
大 连	Dalian	1 152	1 600	1 900
丹 东	Dandong	431	586	773
锦 州	Jinzhou	306	328	426
营 口	Yingkou	205	231	242
盘 锦	Panjin	160	198	260
葫芦岛	Huludao	304	400	460
上 海	**Shanghai**	**7 603**	**8 505**	**9 012**
江 苏	**Jiangsu**	**1 432**	**1 703**	**1 973**
南 通	Nantong	515	629	743
连云港	Lianyungang	502	609	700
盐 城	Yancheng	415	465	530
浙 江	**Zhejiang**	**8 902**	**10 469**	**12 481**
杭 州	Hangzhou	2 776	3 016	3 266
宁 波	Ningbo	1 720	2 010	2 352
温 州	Wenzhou	769	1 052	1 401
嘉 兴	Jiaxing	905	1 117	1 363
绍 兴	Shaoxing	1 028	1 211	1 503
舟 山	Zhoushan	640	825	988
台 州	Taizhou	1 064	1 238	1 608

续表 continued

城　市　City		2004	2005	2006
山　东	**Shandong**	**4 555**		**6 968**
青　岛	Qingdao	1 655		2 449
东　营	Dongying	125		197
烟　台	Yantai	922		1 436
潍　坊	Weifang	594		887
威　海	Weihai	670		981
日　照	Rizhao	393		777
滨　州	Binzhou	198		241
广　东	**Guangdong**	**5 313**	**6 961**	**7 605**
广　州	Guangzhou	656	2 238	2 340
深　圳	Shenzhen	1 014	1 383	1 526
珠　海	Zhuhai	333	421	467
汕　头	Shantou	44	264	424
江　门	Jiangmen	477	512	543
湛　江	Zhanjiang	434	121	137
茂　名	Maoming	15	144	140
惠　州	Huizhou	249	321	380
汕　尾	Shanwei	381	147	149
阳　江	Yangjiang	273	238	239
东　莞	Dongguan	1 006	589	592
中　山	Zhongshan	329	368	404
潮　州	Chaozhou	99	127	146
揭　阳	Jieyang	3	88	118
广　西	**Guangxi**	**619**	**552**	**786**
北　海	Beihai	369	399	440
防城港	Fangchenggang	114	153	170
钦　州	Qinzhou	137		176
海　南	**Hainan**	**563**	**656**	**720**
海　口	Haikou	276	307	335
三　亚	Sanya	287	349	385

注：数据来源于《中国区域经济统计年鉴》。

Note：The data come from "China Statistical Yearbook For Regional Economy".

3-18 沿海城市国际旅游外汇收入
Foreign Exchange Receipts from International Tourism by Coastal Cities

单位：万美元 Unit:10 000 US$

城 市	City	2004	2005	2006
合 计	**Total**	**1 165 235**	**1 420 816**	**1 670 041**
天 津	**Tianjin**	**41 253**	**50 901**	**62 590**
河 北	**Hebei**	**9 508**	**10 373**	**11 829**
唐 山	Tangshan	1 376	1 011	1 247
秦皇岛	Qinhuangdao	8 040	9 257	10 407
沧 州	Cangzhou	92	105	175
辽 宁	**Liaoning**	**40 465**	**47 662**	**57 328**
大 连	Dalian	35 000	40 000	46 500
丹 东	Dandong	2 770	3 680	4 493
锦 州	Jinzhou	975	1 841	3 095
营 口	Yingkou	671	718	943
盘 锦	Panjin	479	747	1 359
葫芦岛	Huludao	570	676	938
上 海	**Shanghai**	**304 124**	**355 588**	**390 399**
江 苏	**Jiangsu**	**14 990**	**20 257**	**26 285**
南 通	Nantong	9 934	13 531	17 496
连云港	Lianyungang	3 520	4 601	6 096
盐 城	Yancheng	1 536	2 125	2 693
浙 江	**Zhejiang**	**97 518**	**143 098**	**176 051**
杭 州	Hangzhou	59 734	75 804	90 870
宁 波	Ningbo	15 428	24 833	33 703
温 州	Wenzhou	7 102	9 097	10 926
嘉 兴	Jiaxing		13 849	16 078
绍 兴	Shaoxing	5 356	6 679	8 890
舟 山	Zhoushan	5 976	7 532	9 127
台 州	Taizhou	3 922	5 304	6 457

续表 continued

城市 City		2004	2005	2006
福 建	**Fujian**	**101 282**	**124 260**	**140 292**
福 州	Fuzhou	23 538	27 266	25 225
厦 门	Xiamen	42 296	55 233	59 083
莆 田	Putian	1 832	2 585	3 968
泉 州	Quanzhou	32 702	37 728	42 888
漳 州	Zhangzhou	914	1 400	9 026
宁 德	Ningde		48	102
山 东	**Shandong**	**46 793**	**64 911**	**84 135**
青 岛	Qingdao	29 182	41 493	54 262
东 营	Dongying	124	77	240
烟 台	Yantai	10 479	13 207	16 590
潍 坊	Weifang	903	1 055	1 293
威 海	Weihai	5 097	7 086	8 929
日 照	Rizhao	1 008	1 908	2 654
滨 州	Binzhou		85	167
广 东	**Guangdong**	**501 147**	**591 036**	**698 188**
广 州	Guangzhou	189 665	228 466	279 728
深 圳	Shenzhen	178 672	200 869	226 550
珠 海	Zhuhai	56 258	70 148	87 389
汕 头	Shantou	6 574	5 950	6 093
江 门	Jiangmen	9 254	10 521	11 554
湛 江	Zhanjiang	1 071	1 483	1 604
茂 名	Maoming	558	676	862
惠 州	Huizhou	8 628	16 549	21 315
汕 尾	Shanwei	389	455	479
阳 江	Yangjiang	794	852	998
东 莞	Dongguan	23 437	28 189	33 180
中 山	Zhongshan	20 688	21 027	21 682
潮 州	Chaozhou	4 614	5 264	6 115
揭 阳	Jieyang	545	587	639
广 西	**Guangxi**	**1 082**	**1 256**	**1 767**
北 海	Beihai	553	642	969
防城港	Fangchenggang	485	545	706
钦 州	Qinzhou	44	69	92
海 南	**Hainan**	**7 074**	**11 474**	**21 177**
海 口	Haikou	2 736	4 023	4 111
三 亚	Sanya	4 338	7 451	17 066

3-19 沿海城市接待入境旅游者人数
Inbound Tourists to Coastal Cities

单位：人次 Unit:person-times

城 市	City	2004	2005	2006
合 计	**Total**	**25 497 922**	**30 422 033**	**34 398 855**
天 津	**Tianjin**	**615 900**	**740 071**	**880 588**
河 北	**Hebei**	**203 206**	**213 106**	**242 238**
唐 山	Tangshan	37 478	30 189	37 387
秦皇岛	Qinhuangdao	162 751	179 217	199 451
沧 州	Cangzhou	2 977	3 700	5 400
辽 宁	**Liaoning**	**682 008**	**814 452**	**979 530**
大 连	Dalian	520 035	600 030	700 032
丹 东	Dandong	86 583	103 494	128 000
锦 州	Jinzhou	29 499	54 952	68 117
营 口	Yingkou	16 400	17 011	21 109
盘 锦	Panjin	8 836	14 464	33 626
葫芦岛	Huludao	20 655	24 501	28 646
上 海	**Shanghai**	**3 854 505**	**4 445 428**	**4 646 303**
江 苏	**Jiangsu**	**184 265**	**245 773**	**295 315**
南 通	Nantong	109 966	151 305	180 617
连云港	Lianyungang	40 298	53 758	68 891
盐 城	Yancheng	34 001	40 710	45 807
浙 江	**Zhejiang**	**2 061 799**	**3 026 261**	**3 680 869**
杭 州	Hangzhou	1 234 016	1 514 582	1 820 420
宁 波	Ningbo	322 417	438 520	542 673
温 州	Wenzhou	172 497	213 419	253 350
嘉 兴	Jiaxing		442 458	527 365
绍 兴	Shaoxing	153 446	201 933	280 303
舟 山	Zhoushan	116 513	140 028	167 668
台 州	Taizhou	62 910	75 321	89 090
福 建	**Fujian**	**1 582 201**	**1 767 137**	**2 093 914**
福 州	Fuzhou	330 779	308 883	315 018
厦 门	Xiamen	620 539	767 727	902 620

续表 continued

城 市 City		2004	2005	2006
莆 田	Putian	94 753	110 665	131 622
泉 州	Quanzhou	500 783	535 381	555 705
漳 州	Zhangzhou	35 347	42 089	185 500
宁 德	Ningde		2 392	3 449
山 东	**Shandong**	**873 174**	**1 138 308**	**1 428 313**
青 岛	Qingdao	522 498	684 407	854 462
东 营	Dongying	1 802	1 251	4 692
烟 台	Yantai	149 109	185 124	239 690
潍 坊	Weifang	30 848	35 840	43 052
威 海	Weihai	121 739	168 350	204 866
日 照	Rizhao	47 178	59 536	75 665
滨 州	Binzhou		3 800	5 886
广 东	**Guangdong**	**15 134 285**	**17 610 148**	**19 542 314**
广 州	Guangzhou	4 371 493	5 103 080	5 642 267
深 圳	Shenzhen	5 597 637	6 164 399	7 001 378
珠 海	Zhuhai	1 932 425	2 376 707	2 779 110
汕 头	Shantou	216 666	201 953	163 161
江 门	Jiangmen	287 270	412 999	346 021
湛 江	Zhanjiang	28 489	33 284	38 896
茂 名	Maoming	15 209	22 460	21 852
惠 州	Huizhou	318 644	798 366	927 891
汕 尾	Shanwei	20 527	22 420	23 716
阳 江	Yangjiang	44 425	55 941	61 800
东 莞	Dongguan	1 496 383	1 594 207	1 567 609
中 山	Zhongshan	621 630	666 100	692 744
潮 州	Chaozhou	158 600	130 232	241 398
揭 阳	Jieyang	24 887	28 000	34 471
广 西	**Guangxi**	**59 071**	**69 091**	**89 548**
北 海	Beihai	25 334	30 228	40 245
防城港	Fangchenggang	32 200	35 763	45 286
钦 州	Qinzhou	1 537	3 100	4 017
海 南	**Hainan**	**247 508**	**352 258**	**519 923**
海 口	Haikou	101 236	140 278	131 105
三 亚	Sanya	146 272	211 980	388 818

3-20 沿海城市接待入境旅游者情况(2006年)

Breakdown of Inbound Tourists by Coastal Cities, 2006

单位：人次，人天 Unit:person-time, night

城市 City	外国人 Foreigners		港澳同胞 Hongkong & Macau		台湾同胞 Taiwan Province	
	人次数 Arrivals	人天数 Nights	人次数 Arrivals	人天数 Nights	人次数 Arrivals	人天数 Nights
合计 Total	**16 134 328**	**49 195 619**	**13 857 278**	**31 240 501**	**4 407 249**	**11 284 750**
天津 Tianjin	**811 691**	**2774 352**	**30 815**	**247 413**	**38 082**	**243 592**
河北 Hebei	**228 865**	**759 628**	**7 501**	**20 214**	**5 872**	**7 649**
唐山 Tangshan	35 382	83 097	935	2 010	1 070	1 720
秦皇岛 Qinhuangdao	190 673	670 266	4 766	14 189	4 012	4 189
沧州 Cangzhou	2 810	6 265	1 800	4 015	790	1 740
辽宁 Liaoning	**844 927**	**2 692 030**	**74 321**	**194 563**	**60 282**	**195 907**
大连 Dalian	616 134	2 204 404	34 346	117 461	49 552	173 928
丹东 Dandong	126 129	241 225	1 235	1 503	636	723
锦州 Jinzhou	49 141	121 972	14 502	34 391	4 474	11 338
营口 Yingkou	20 778	50 411	232	418	99	240
盘锦 Panjin	14 754	40 162	16 899	29 783	1 973	3 714
葫芦岛 Huludao	17 991	33 856	7 107	11 007	3 548	5 964
上海 Shanghai	**3 997 979**	**14 501 002**	**331 319**	**1076 592**	**317 005**	**1 157 625**
江苏 Jiangsu	**244 511**	**1 520 931**	**15 268**	**135 746**	**35 536**	**270 561**
南通 Nantong	158 549	929 068	9 132	73 235	12 936	122 536
连云港 Lianyungang	59 726	376 206	3 623	33 347	5 542	71 799
盐城 Yancheng	26 236	215 657	2 513	29 164	17 058	76 226

续表 continued

城 市 City	外国人 Foreigners		港澳同胞 Hongkong & Macau		台湾同胞 Taiwan Province	
	人次数 Arrivals	人天数 Nights	人次数 Arrivals	人天数 Nights	人次数 Arrivals	人天数 Nights
浙 江 Zhejiang	**2 414 813**	**6 035 440**	**601 100**	**1 354 256**	**664 956**	**1 672 006**
杭 州 Hangzhou	1 236 735	3 120 586	258 710	674 030	324 975	880 768
宁 波 Ningbo	333 282	1 112 446	98 696	260 763	110 695	359 505
温 州 Wenzhou	200 510	457 341	27 313	51 768	25 527	55 939
嘉 兴 Jiaxing	289 065	542 805	120 624	147 228	117 676	142 858
绍 兴 Shaoxing	194 741	326 330	51 079	73 577	34 483	63 711
舟 山 Zhoushan	106 036	312 357	25 085	61 020	36 547	99 427
台 州 Taizhou	54 444	163 575	19 593	85 870	15 053	69 798
福 建 Fujian	**716 175**	**2 920 537**	**696 968**	**3 350 956**	**680 771**	**2 128 201**
福 州 Fuzhou	196 513	869 639	41 989	216 809	76 516	348 203
厦 门 Xiamen	407 962	1 779 664	110 847	350 128	383 811	989 880
莆 田 Putian	8 066	20 774	13 958	46 435	109 598	160 540
泉 州 Quanzhou	50 778	99 009	451 406	2 509 816	53 521	465 046
漳 州 Zhangzhou	50 967	147 804	78 331	227 160	56 202	162 986
宁 德 Ningde	1 889	3 647	437	608	1 123	1 546
山 东 Shandong	**1 259 330**	**3 925 331**	**88 818**	**199 047**	**80 165**	**244 788**
青 岛 Qingdao	751 403	2 562 098	59 192	138 593	43 867	106 694
东 营 Dongying	2 630	10 163	1 260	2 390	802	1 495
烟 台 Yantai	196 756	716 842	21 479	45 692	21 455	104 234
潍 坊 Weifang	33 178	57 966	4 965	8 508	4 909	8 787
威 海 Weihai	194 172	403 717	1 738	3 550	8 956	23 319
日 照 Rizhao	75 498	165 255	72	134	95	113
滨 州 Binzhou	5 693	9 290	112	180	81	146

续表 continued

城市 City	外国人 Foreigners		港澳同胞 Hongkong & Macau		台湾同胞 Taiwan Province	
	人次数 Arrivals	人天数 Nights	人次数 Arrivals	人天数 Nights	人次数 Arrivals	人天数 Nights
广 东 Guangdong	**5 153 659**	**12 826 897**	**11 914 537**	**24 469 960**	**2 474 118**	**5 255 037**
广 州 Guangzhou	2 075 341	5 216 990	3 130 104	7 573 686	436 822	1 169 736
深 圳 Shenzhen	1 372 852	3 355 000	5 097 109	9 693 076	531 417	1 187 474
珠 海 Zhuhai	433 410	949 669	1 460 768	2 301 966	884 932	1 736 678
汕 头 Shantou	83 503	275 560	63 173	144 270	16 485	46 158
江 门 Jiangmen	50 576	854 619	228 159	538 733	67 286	134 214
湛 江 Zhanjiang	19 603	70 268	17 539	40 104	1 754	2 188
茂 名 Maoming	4 558	14 783	13 882	42 860	3 412	10 748
惠 州 Huizhou	209 427	573 820	599 095	1 241 508	119 369	336 185
汕 尾 Shanwei	291	757	22 671	40 807	754	1 960
阳 江 Yangjiang	2 310	4 620	44 490	78 870	15 000	27 000
东 莞 Dongguan	727 500	1 156 725	555 457	1 270 272	284 652	395 666
中 山 Zhongshan	151 310	304 510	455 976	1 040 145	85 458	157 952
潮 州 Chaozhou	17 900	42 960	204 661	431 999	18 837	34 095
揭 阳 Jieyang	5 078	6 616	21 453	31 664	7 940	14 983
广 西 Guangxi	**70 544**	**91 287**	**12 704**	**18 626**	**6 300**	**9 979**
北 海 Beihai	25 789	38 481	9 824	14 797	4 632	7 429
防城港 Fangchenggang	43 080	50 749	1 647	2 004	559	777
钦 州 Qinzhou	1 675	2 057	1 233	1 825	1 109	1 773
海 南 Hainan	**391 834**	**1 148 184**	**83 927**	**173 128**	**44 162**	**99 405**
海 口 Haikou	80 124	165 802	25 457	45 855	25 524	54 937
三 亚 Sanya	311 710	982 382	58 470	127 273	18 638	44 468

主要统计指标解释

1.海洋捕捞产量 凡是从海洋里捕捞的天然生长的水产品产量为捕捞产量。

2. 海水养殖产量 凡是从人工投放苗种或天然纳苗并进行人工饲养管理的海水养殖水域中捕捞的水产品产量为海水养殖产量。

3. 远洋捕捞产量 由各远洋渔业企业和各生产单位按我国远洋渔业项目管理办法组织的远洋渔船(队)在非我国管辖海域(外国专属经济区水域或公海)捕捞的水产品产量。中外合资、合作渔船捕捞的水产品只统计按协议应属于中方所有的部分。

4.海滨砂矿 包括在砂质海岸或近岸海底开采的金属砂矿和非金属砂矿,主要品种有:铁矿砂、锡矿砂、砂金和稀有金属砂矿(金红石、钛铁矿、锆英石和独居石等)、金刚石砂矿、砾以及非金属矿(建筑材料)等。

5.原油产量 是按净原油量来计算的,能直接用于销售和生产自用的原油量。目前海洋石油系统原油产量计算方法采用倒算法。

原油产量=销售量+期末库存量-期初库存量+海上平台及陆地终端处理厂自用量。

6.天然气产量 指进入集输管网的销售量和就地利用的全部气量。

天然气产量=外输(销)量+企业自用量

7.海洋原油出口量 指销往国外的产品数量。

8.海洋原油出口创汇额 指产品销往国外的归中方所有的全部外汇收入。以美元或万美元表示。

9.造船综合吨 等于以计量单位载重吨和满载排水量吨的民用船舶的吨位数之和。

10.货运量 指经船舶实际运送的货物重量,按到达量统计。

11 货物周转量 指实际运送的货物与其运送距离的乘积。

12.集装箱运量 既包含货重,也包含箱重。箱重系指承运租用的空箱重量凡有运费收入的空箱,其重量应统计为运量,按空箱 1 吨为货运量 1 吨计算;若无收入,所承运的空箱一律不作运量统计。

13.旅客周转量 指实际运送的旅客人数与其运送距离的乘积。

14.国际旅游外汇收入 入境旅游者在中国(大陆)境内旅行、游览过程中用于交通、参观游览、住宿、餐饮、购物、娱乐等的全部花费。

15.接待人次数 指报告期内我国接待游客人数。游客按出游地分为国际游客(即海外游客)和国内游客,按出游时间分为旅游者(过夜游客)和一日游游客(不过夜游客)。

16.接待人天数 指过夜旅游者的停留天数。

17.外国人 指外国国籍的人,加入外籍的中国血统华人也计入外国人。

18.港澳台同胞 指居住在我国香港特别行政区、澳门特别行政区和台湾省的中国同胞。

Explanatory Notes on Main Statistical Indicators

1.Marine Catches refers to the output of the naturally growing aquatic products caught from the sea.

2.Mariculture Production refers to the output of aquatic products whose young are artificially released or naturally collected, and raised and managed artificially, and which are caught from the waters of mariculture.

3.Deep–Sea Fishing Production refers to the output of aquatic products caught in the non-Chinese jurisdictional sea areas (foreign EEZ or high sea) by the distant fishing vessels (fleet) organized by various distant fishing businesses and production units according to the management measures of the China distant fishing projects. The aquatic products caught by the Chinese-foreign joint ventures' and cooperative fishing vessels are counted only for the part owned by the Chinese side according to the agreement.

4.Beach Placers include the metallic and non-metallic placers exploited on the sandy coast or nearshore seabed. The main species of beach placer are iron ore sand, tin ore sand, placer gold and rare metals placers (rutile, ilmenite, zircon and monazite), diamond placer and gravel as well as nonmetallic minerals (building materiels).

5.Output of Crude Oil is calculated on the basis of the net amount of crude oil, i.e., the amount of crude oil that may be directly used for sale and for the production itself.

Output of crude oil = Volume of sales + Reserves at the end of the period-Reserves at the beginning of the period +Amount for self-use on the platforms and in the terminal processing plants on land.

6.Output of Natural Gas refers to the total gas volume of sales volume entering the oil collecting and transport pipeline network and that used locally.

Output of natural gas = Volume of sales or transport to other areas + Volume used by the enterprise itself

7.Export Volume of Offshore Crude Oil refers to the amount of products for sale abroad.

8.Foreign Exchange Earnings of Offshore Crude Oil refers to the total foreign exchange income from oil (gas) products for sale abroad which is owned by the Chinese side.

9.Comprehensive Tonnages of Shipbuilding refers to the sum of tonnage of civilian vessels with the deadweight capacity and full-load displacement as measured.

10.Freight Traffic refers to the weight of cargoes actually transported by vessels, which is counted according to the volume of arrival.

11.Cargoes Turnover Volume refers to the product of the actually transported cargoes and the transport distance.

12.Freight Volume of Containers includes the weight of both cargoes and container boxes. The weight of container boxes refers to the weight of empty containers rented for transport or having freight income and should be included in the freight volume, one ton of empty boxes equalling to one ton of freight volume. The empty boxes which have no income for transportation are not included in the freight volume.

13.Passenger Turnover Volume refers to the product of the number of passengers actually transported and the shipping distance.

14.International Tourism (Foreign Exchange) Receipts refers to the total expenditure made by inbound tourists within the territory of China (the mainland) in their course of travel on transport, tours and sightseeing, lodging, food and beverage, shopping, entertainment and etc.

15.Number of Person–Times Received refers to the number of tourists received by China in the period reported. Visitors are divided into the international visitors (i.e., inbound tourists) and domestic visitors by origin of the travel, and tourists (overnight tourists) and same-day tourists (non-overnight tourists) by their length of stay.

16.Number of the Days of Stay refers to the number of the days of stay of overnight tourists.

17.Foreigners refer to the people with foreign nationality, including foreign nationals of Chinese descent.

18.Compatriots from Hong Kong, Macau and Taiwan Province refer to the Chinese compatriots living in the Hong Kong Special Administrative Region, the Macau Special Administrative Region and Taiwan Province.

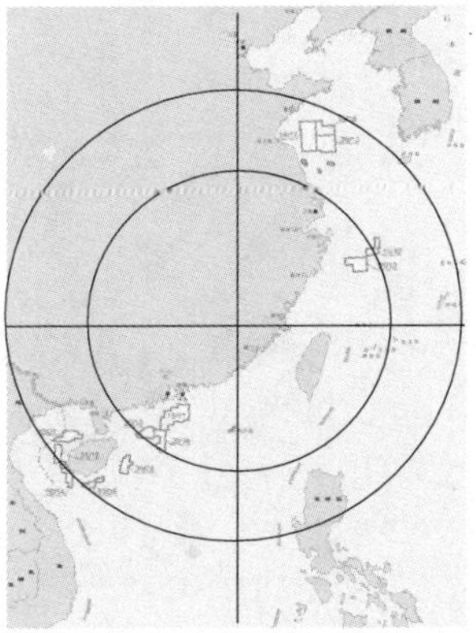

4

主要海洋产业生产能力

Production Capacity of Major Marine Industries

4-1 浅海滩涂海湾可养殖面积
Seashore Land Area for Cultivation

单位: 千公顷 Unit: 1 000 hm^2

地区 Region		海水可养殖面积 Cultivatable Marine Area	浅海 Shallow Sea	滩涂 Tidal Flat	港湾 Harbor
全国总计*	**National Total**	**2 600.11**	**1 622.56**	**797.00**	**180.55**
天津	Tianjin	18.49	10.00	8.49	
河北	Hebei	111.37	49.66	61.70	
辽宁	Liaoning	725.84	590.44	92.45	42.95
上海	Shanghai	3.22		3.22	
江苏	Jiangsu	139.00	7.87	130.96	0.17
浙江	Zhejiang	101.46	36.30	57.39	7.77
福建	Fujian	184.94	77.39	100.76	6.79
山东	Shandong	358.21	131.68	173.41	53.12
广东	Guangdong	835.67	664.00	120.00	51.67
广西	Guangxi	31.95	6.78	22.09	3.08
海南	Hainan	89.52	48.43	26.09	15.00

注：*全国总计中包含北京数据。

Note: * Data for Beijing are included in the national total.

4-2 沿海地区海水养殖面积（2006年）

Mariculture Area by Coastal Regions, 2006

单位：公顷 Unit:hm²

地 区 Region	海水养殖面积 Mariculture Area
全国总计 National Total	**1 774 119**
天 津 Tianjin	5 511
河 北 Hebei	102 302
辽 宁 Liaoning	485 800
上 海 Shanghai	80
江 苏 Jiangsu	180 680
浙 江 Zhejiang	109 055
福 建 Fujian	152 298
山 东 Shandong	420 258
广 东 Guangdong	234 635
广 西 Guangxi	64 336
海 南 Hainan	19 164

4-3 海洋油田生产井情况（2006年）
Survey of Offshore Oilfield Production Wells, 2006

单位：口 Unit:number

地 区 Region	合 计 Total	采油井 Oil Wells	采气井 Gas Wells	注水井 Injection Wells	其他井 Others
合 计 Total	**2 632**	**2 036**	**153**	**429**	**14**
天 津 Tianjin	1 194	927	74	193	
河 北 Hebei	495	366		129	
辽 宁 Liaoning	245	180	20	31	14
上 海 Shanghai	22	8	14		
山 东 Shandong	352	280	4	68	
广 东 Guangdong	324	275	41	8	

4-4 海洋石油勘探情况（2006年）
Work Volume of Offshore Oil Exploration, 2006

地 区 Region	地震测线 Seismic Line		钻井（口） Drilling (Well)	
	二维（公里） two dimensions (km)	三维（平方公里） three dimensions (km^2)	预探井 Wildcat Wells	评价井 Appraisal Wells
合 计 Total	**20 819**	**7 638**	**93**	**19**
天 津 Tianjin	0	2 629	30	7
其中：合作 Including: Cooperative	0	365	4	0
河 北 Hebei	0	415	24	7
辽 宁 Liaoning	0	250	6	2
上 海 Shanghai	6 646	550	0	0
其中：合作 Including: Cooperative	6 646	550	0	0
山 东 Shandong	0	0	21	0
广 东 Guangdong	14 173	3 794	12	3
其中：合作 Including: Cooperative	0	1 430	3	0

4-5 沿海地区盐田面积和海盐生产能力

Salt Pan Area and Sea Salt Production Capacity by Coastal Regions

地区 Region	盐田总面积（公顷） Total Area of Salt Pan (hm^2)		生产面积（公顷） Production Area (hm^2)		年末海盐生产能力（万吨） Year-End Capacity of Sea Salt Production (10 000 tons)	
	2005	2006	2005	2006	2005	2006
合 计 Total	**402 989**	**443 025**	**335 151**	**356 918**	**3 226.78**	**4 165.21**
天 津 Tianjin	33 779	33 649	32 582	32 446	199.70	197.00
河 北 Hebei	87 204	81 393	78 707	77 582	537.20	535.00
辽 宁 Liaoning	55 852	56 274*	49 589	46 262*	295.60	283.63*
江 苏 Jiangsu	73 023	68 464	46 553	42 690	209.70	193.20
浙 江 Zhejiang	6 597	5 193	5 311	4 366	29.49	27.46
福 建 Fujian	6 648	6 894	5 975	5 610	54.26	49.59
山 东 Shandong	122 273	173 721	105 221	136 124	1 844.93	2 822.23
广 东 Guangdong	11 003*	10 484*	6 188*	6 603*	20.25*	20.22*
广 西 Guangxi	3 022	3 017	1 750	1 714	12.05	12.05
海 南 Hainan	3 588	3 936	3 275	3 521	23.60	24.83

注：*为中国盐业总公司数据。

Note: * is the data for the China Salt Industry Corporation.

4-6 沿海地区风能发电能力（2005年）

Wind Power Production by Coastal Regions, 2005

单位：万千瓦　Unit: 10 000 kW

地　区 Region	风能年发电能力 Annual Wind Power Generation Capacity
合　计	**41.59**
Total	
河　北	4.80
Hebei	
辽　宁	12.56
Liaoning	
上　海	2.44
Shanghai	
浙　江	3.72
Zhejiang	
福　建	5.20
Fujian	
山　东	3.66
Shandong	
广　东	8.34
Guangdong	
海　南	0.87
Hainan	

4-7 主要潮汐电站分布情况（2006年）
Distribution of Major Tidal Power Stations, 2006

电站名称 Name	分布位置 Location
江厦潮汐试验电站	浙江省乐清湾
Jiangxia Experimental Tidal Power Station	Yueqing Bay,Zhejiang Province
海山潮汐电站	浙江省
Haishan Tidal Power Station	Zhejiang Province
沙山潮汐电站	浙江省
Shashan Tidal Power Station	Zhejiang Province
岳浦潮汐电站	浙江省象山县
Yuepu Tidal Power Station	Xiangshan County,Zhejiang Province
白沙口潮汐电站	山东省乳山县
Baishakou Tidal Power Station	Rushan County,Shandong Province
浏河潮汐电站	江苏省太仓县
Liuhe Tidal Power Station	Taicang County,Jiangsu Province
果子山潮汐电站	广西钦州湾
Guozishan Tidal Power Station	Qinzhou Bay,Guangxi
幸福洋潮汐电站	福建省平潭县
Xingfuyang Power Station	Pingtan County,Fujian Province

4-8 主要海上活动船舶（2006年）

Major Vessels Operating on the Sea, 2006

类别 Type	艘数 (艘) Number of Vessels	总吨 (万吨位) grt (10 000 tons)	净载重量 (万吨) Net Weight Tonnage (10 000 tons)	载客量 (客位) Passenger Spaces	总功率 (千瓦) Total Power (kW)
一、海洋生产用船 Vessels for Marine Production					
海洋渔业船舶 Marine Fishing Vessels	**283 607**	**6 205 163**			**14 203 414**
远洋渔船 Ocean-going Fishing Vessels	1 599	654 559			875 322
海洋油气船舶 Offshore Oil and Gas Vessels	**147**	**12.00**			**76 444**
钻井船 Drilling Vessels	15	12.00			76 444
物探船 Physical Exploration Vessels	10				
其他 Others	122				
海洋运输船舶 Maritime Transport Vessels	**11 491**	**4 185.35**	**6 083.39**	**156 636**	**20 180 169**
二、海洋科研用船 Vessels for Marine Scientific Research					
海洋地质勘探船 Marine Geology Survey Vessels	**7**				
海洋调查船 Marine Survey Vessels	**9**	**3.40**	**1.60**	**635**	**45 773**
中国科学院 Chinese Academy of Sciences	5	0.80	0.25	317	20 761
国家海洋局 State Oceanic Administration	4	2.60	1.35	318	25 012
三、海洋执法用船 Vessels for Marine Law Enforcement					
#海监船 Marine Monitoring Vessels	**98**	**3.92**			**97 860**

4-9 沿海主要港口生产用码头泊位（截至2006年底）

Berths for Productive Use at Principal Seaports (by the End of 2006)

单位：米，个 Unit:m, number

港口	Seaport	码头长度 Length of Quay Line	泊位个数 Number of Berths	#万吨级 10 000 Tonnage Class
总计	**Total**	**382 492**	**3 291**	**883**
丹东	Dandong	2 558	15	9
大连	Dalian	31 112	191	73
营口	Yingkou	7 846	36	25
锦州	Jinzhou	3 642	15	13
秦皇岛	Qinhuangdao	12 248	52	42
黄骅	Huanghua	2 946	14	10
唐山	Tangshan	4 444	18	18
天津	Tianjin	22 871	119	65
龙口	Longkou	3 577	22	11
蓬莱	Penglai	1 010	14	
烟台	Yantai	7 961	46	25
威海	Weihai	1 554	12	4
青岛	Qingdao	13 380	54	42
日照	Rizhao	7 838	31	27
连云港	Lianyungang	7 245	35	29
射阳	Sheyang	757	17	
上海	Shanghai	22 194	121	80
宁波	Ningbo	31 415	304	60
舟山	Zhoushan	21 010	403	15
嘉兴	Jiaxing	3 567	30	10
台州	Taizhou	4 536	57	4
温州	Wenzhou	10 394	146	9
福州	Fuzhou	12 177	133	26
湄州湾	Meizhou Bay	1 876	27	2
泉州	Quanzhou	7 703	59	12
厦门	Xiamen	8 502	111	37

续表 continued

港口 Seaport		码头长度 Length of Quay Line	泊位个数 Number of Berths	#万吨级 10 000 Tonnage Class
漳州	Zhangzhou	1 029	13	
汕头	Shantou	8 662	82	16
汕尾	Shanwei	935	10	–
惠州	Huizhou	5 848	25	10
深圳	Shenzhen	23 895	131	57
太平	Taiping	1 872	16	5
广州	Guangzhou	39 886	474	57
中山	Zhongshan	1 703	28	
珠海	Zhuhai	8 580	101	9
江门	Jiangmen	6 968	95	
新会	Xinhui	640	10	
阳江	Yangjiang	834	4	4
茂名	Maoming	1 649	14	6
湛江	Zhanjiang	9 011	71	29
海安	Haian	564	14	1
北海	Beihai	2 458	22	4
钦州	Qinzhou	3 568	24	9
防城	Fangcheng	4 070	25	14
洋浦	Yangpu	1 021	5	5
海口	Haikou	2 922	32	3
八所	Basuo	1 412	8	6
三亚	Sanya	602	5	

注：厦门港统计范围包括原厦门港及招银港区、后石港区、石码港区；烟台港统计范围包括原烟台港及原蓬莱东港区；京唐更名为唐山港，统计范围包括原京塘港及曹妃甸港区。

Note：The Xiamen Harbor statistics covers the original Xiamen Harbor and Zhaoyin Harbor area, Houshi Harbor area and Shima Harbor area.The Yantai Harbor statistics covers the original Yantai Harbor and the original Donggang Harbor area of Penglai;Jingtang Harbor changes its name into Tangshan Harbor,and its statistics cover the original Jingtang Harbor and Caofeidian Harbor area.

4-10 主要沿海城市星级饭店基本情况（2006年）

Star-rated Hotels and Occupancies in Major Coastal Cities, 2006

城 市 City	饭店数（座） Number of Hotels	客房数（间） Number of Rooms	床位数（张） Number of Beds	客房出租率 (%) Room Occupancy (%)
合 计 Total	**2 646**	**381 899**	**674 953**	**62.49**
天 津 Tianjin	112	15 073	27 002	57.32
秦皇岛 Qinhuangdao	65	8 638	16 734	50.12
大 连 Dalian	164	19 867	35 042	62.45
上 海 Shanghai	317	58 900	97 017	63.67
南 通 Nantong	52	6 184	14 746	64.45
连云港 Lianyungang	74	6 020	15 665	54.36
杭 州 Hangzhou	223	32 091	58 236	66.15
宁 波 Ningbo	194	18 506	32 030	63.55
温 州 Wenzhou	79	9 581	16 353	69.95
福 州 Fuzhou	85	13 118	20 783	57.80
厦 门 Xiamen	63	9 863	16 184	61.31
泉 州 Quanzhou	77	8 542	13 777	63.15
漳 州 Zhangzhou	26	2 413	4 013	36.99
青 岛 Qingdao	125	17 264	32 898	71.32
烟 台 Yantai	85	10 684	21 168	67.51
威 海 Weihai	72	7 493	14 940	62.15
广 州 Guangzhou	232	43 596	77 637	62.14
深 圳 Shenzhen	188	34 356	54 438	65.37
珠 海 Zhuhai	85	13 527	22 675	64.68
汕 头 Shantou	42	7 074	12 225	57.06
湛 江 Zhanjiang	43	4 429	8 120	61.34
中 山 Zhongshan	37	4 139	6 784	50.59
北 海 Beihai	41	5 247	9 938	42.56
海 口 Haikou	77	11 743	21 332	57.88
三 亚 Sanya	88	13 551	25 216	58.60

4-11 沿海地区旅行社单位数（2006年）
Number of Travel Agencies by Coastal Regions, 2006

单位：家 Unit:number

地 区 Region		旅行社总数 Number of Travel Agencies	国际旅行社 International Travel Agencies	国内旅行社 Domestic Travel Agencies
合 计	**Total**	**9 192**	**747**	**8 445**
天 津	Tianjin	245	27	218
河 北	Hebei	941	43	898
辽 宁	Liaoning	1 029	77	952
上 海	Shanghai	805	51	754
江 苏	Jiangsu	1 403	78	1 325
浙 江	Zhejiang	1 169	54	1 115
福 建	Fujian	565	44	521
山 东	Shandong	1 640	86	1 554
广 东	Guangdong	913	187	726
广 西	Guangxi	335	61	274
海 南	Hainan	147	39	108

主要统计指标解释

1.海水可养殖面积 指利用滩涂、浅海、港湾进行鱼、虾、蟹、贝、藻等海水经济动植物的人工养殖的水面面积。

2.浅海养殖 指在可养殖的浅海中进行海水经济动植物养殖。

3.港湾养殖 指利用港、湾,或在海边、河口附近的滩涂、洼地拦闸筑堤进行海水养殖。

4.海水养殖面积 是指利用海上、滩涂、陆基进行鱼、甲壳类(虾、蟹)、贝、藻等海水经济动植物的人工养殖的水面面积。在报告期内无论是否全部收获或尚未收获其产品,均应统计在海水养殖面积中。但有些滩涂、水面不投放苗种或投放少量苗种,只进行一般管理的,不统计为养殖面积。

5.盐田总面积 指盐田占有的全部面积。包括储卤、蒸发、保卤、结晶面积、滩内的沟、壕、池、埝、滩坨等面积及滩外的沟、壕、公路及杂地面积。

6.生产面积 指直接提供给海盐生产的面积,包括结晶面积、蒸发面积、保卤面积,滩内的沟、壕、池、埝面积及滩坨面积。

7. 年末海盐生产能力 指年末企业生产原盐的全部设备的综合平衡能力。海盐生产露天作业,受天气影响,因而计算生产能力时,成熟滩田按10年实际平均单位生产面积产量乘以本年成熟滩田生产面积而得,新滩田按设计能力及滩田成熟程度可能达到的产量计算。

8.海洋渔业船舶 是指配置机器作为动力的从事海洋渔业生产和辅助渔业生产的船舶。

9. 远洋渔船 按我国远洋渔业项目管理办法在非我国管辖海域(外国专属经济区水域或公海)进行常年或季节性生产的渔船。

10.泊位个数 是指设有系靠船舶设施,在同一时间内可供靠泊最大吨级船舶的艘数。即可靠泊一艘船舶,则计为一个泊位,余类推。泊位分码头泊位和浮筒泊位。

11.客房数 指饭店实际可用于接待旅游者的房间数。

12.床位数 指饭店实际可用于接待旅游者的床位数。

Explanatory Notes on Main Statistical Indicators

1.Marine Cultivatable Area refers to the water area in tidal flat, shallow sea and bay that is used to breed marine economic animals and plants, such as fish, shrimp, crab, shellfish, alga and so on.

2.Shallow Sea Cultivation refers to the breeding of marine economic animals and plants in the cultivatable shallow sea.

3.Harbor Cultivation refers to the marine cultivation conducted in harbors, bays, or the tidal flat or marshes around seaside and bayou by blocking the gate and banking up the dam.

4.Mariculture Area refers to the area of the water surface where seawater economic animals and plants, such as crustacean (shrimp, crab), shellfish and algae, are cultivated at sea, on tidal flat and land. Whether or not all the products in the area have been harvested or the products have not been harvested yet in the period covered by the report, the area is included in the Mariculture Area. But some tidal flats and water surfaces where none or a small amount of the young have been released and only general management is carried out are not included in the Mariculture Area.

5.Total Area of Salt Pans refers to the total area covered by salt pans, including the area for brine storage, evaporation, brine preservation, and crystallization, the area of ditches, moats, pondsand banks within the beach as well as beach mounds, and ditches, moats, highway beyond the beach as well as the area of miscellanneous lands.

6.Area of Salt Pan Production refers to the area directly provided for sea-salt production, including the area for crystallization, evaporation and brine preservation, the area of ditches, moats, ponds and banks within the beach as well as the area of beach mounds.

7.Year–End Capacity of Crude Salt Production refers to the integrated and balanced capacity of all equipment of the enterprise used for crude salt production at the end of the year. As sea salt production is an open-air operation, which is subject to the effect of weather, the production capacity of a matured salt pan is calculated at the production of the actual average unit production area in ten years times the production area of the matured salt pan in the current year. The production capacity of new salt pans is calculated at the production that may be reached in the light of the designed capacity and the level of maturity of the salt pan.

8.Marine Fishing Vessels refer to the vessels equipped with machines as motive power and going for marine fishery production and auxiliary fishery production.

9.Deep Sea Fishing Vessels refer to the fishing vessels which carry out production all the year round or seasonally in the non-Chinese jurisdictional, sea areas (foreign EEZ or high sea) according to the China Deep-sea Fishing Projects Planagement Measures.

10.Number of Berths refers to the spaces equipped with facilities for docking ships and the number of ships of the maximum tonnage that may dock or anchor in them. A space for a ship to dock is counted as one berth and the rest are reasoned out by analogy. Berths are divided into wharf berths and buoy berths.

11.Number of Rooms refers to the number of guests rooms actually used by the hotels receiving tourists.

12.Number of Beds refers to the number of beds actually used by the hotels receiving tourists.

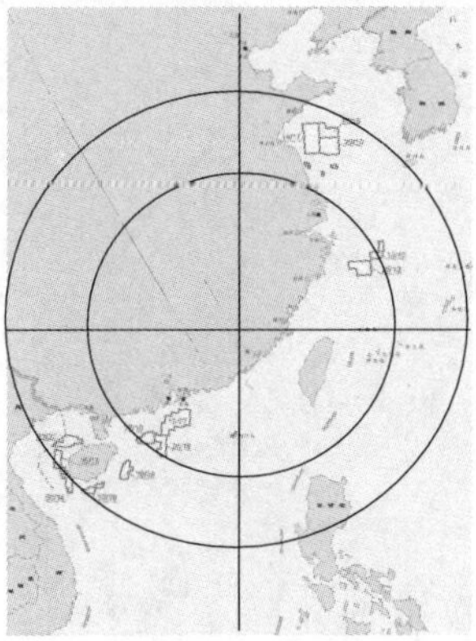

5

涉海就业

Ocean-Related Employment

5-1 沿海地区涉海就业人员情况
Ocean-Related Employed Personnel by Coastal Regions

地 区 Region	2001 总数(万人) Total Number (10 000 persons)	2001 占地区就业人员比重(%) Proportion in the Regional Employed Personnel(%)	2005 总数(万人) Total Number (10 000 persons)	2005 占地区就业人员比重(%) Proportion in the Regional Employed Personnel(%)	2006 总数(万人) Total Number (10 000 persons)
合 计 Total	**2 107.6**	**8.1**	**2 780.8**	**9.7**	**2 960.3**
天 津 Tianjin	106. 4	25. 9	140. 4	32. 9	149. 4
河 北 Hebei	58. 0	1. 7	76. 5	2. 2	81. 5
辽 宁 Liaoning	196. 0	10. 7	258. 6	13. 1	275. 3
上 海 Shanghai	127. 5	18. 4	168. 2	19. 7	179. 1
江 苏 Jiangsu	116. 9	3. 3	154. 2	4. 0	164. 2
浙 江 Zhejiang	256. 4	9. 2	338. 3	10. 6	360. 1
福 建 Fujian	259. 7	15. 5	342. 7	18. 3	364. 8
山 东 Shandong	319. 9	6. 8	422. 1	8. 3	449. 3
广 东 Guangdong	505. 3	12. 8	666. 7	14. 2	709. 7
广 西 Guangxi	68. 9	2. 7	90. 9	3. 4	96. 8
海 南 Hainan	80. 6	23. 7	106. 3	28. 2	113. 2
其 他* Others	12. 0		15. 8		16. 9

注:2005年和2006年为推算数据；其他为非沿海地区涉海就业人员数。

Note:The data for 2005 and 2006 are the calculated ones;Others refer to the number of ocean-related employed persons in the non-coastal regions.

5-2 全国主要海洋产业就业人员情况
Employed Personnel in the Major Marine Industries Throughout the Country

单位：万人 Unit: 10 000 persons

海洋产业 Marine Industry	2001	2005	2006
合计 Total	**719.1**	**949.2**	**1 006.7**
海洋渔业及相关产业 Marine Fishery and the Related Industries	348.3	459.8	487.6
海洋石油和天然气 Offshore Oil and Natural Gas	12.4	16.4	17.4
海滨砂矿 Beach Placer	1.0	1.3	1.4
海洋盐业 Sea Salt Industry	15.0	19.9	21.0
海洋化工 Marine Chemical	16.1	21.3	22.5
海洋生物医药 Marine Biological Pharmaceatical	0.6	0.8	0.8
海洋电力和海水利用 Marine Electric Power and Seawater Utilization	0.7	0.9	1.0
海洋船舶工业 Marine Shipbuilding Industry	20.6	27.2	28.8
海洋工程建筑 Marine Engineering Architecture	38.8	51.2	54.3
海洋交通运输 Maritime Transportation	50.8	67.1	71.1
滨海旅游 Coastal Tourism	78.3	103.4	109.6
其他海洋产业 Other Marine Industries	136.5	180.0	191.2

注:2005年和2006年为推算数据。

Note:The data for 2005 and 2006 are the caculated ones.

5-3 全国主要海洋产业就业人员年龄构成(2001年)

Age Composition of Employees at the End of the Year in the Major Marine Industries, 2001

海洋产业 Marine Industry	就业人数（万人） Number of the Employed(10 000 persons)	35岁以下 Below 35 in age	35岁以下就业人员比重(%) Proportion of the Employed Below 35 in Age(%)
合计 Total	**719.1**	**395.2**	**55.0**
海洋渔业及相关产业 Marine Fishery and the Related Industries	348.3	183.3	52.6
海洋石油和天然气 Offshore Oil and Natural Gas	12.4	12.2	98.4
海滨砂矿 Beach Placer	1.0	0.5	50.0
海洋盐业 Sea Salt Industry	15.0	6.4	42.7
海洋化工 Marine Chemical	16.1	8.9	55.3
海洋生物医药 Marine Biological Pharmaceatical	0.6	0.4	66.7
海洋电力和海水利用 Marine Electric Power and Seawater Utilization	0.7	0.4	57.1
海洋船舶工业 Marine Shipbuilding Industry	20.6	12.6	61.2
海洋工程建筑 Marine Engineering Architecture	38.8	17.5	45.1
海洋交通运输 Maritime Transportation	50.8	22.9	45.1
滨海旅游 Coastal Tourism	78.3	54.0	69.0
其他海洋产业 Other Marine Industries	136.5	76.1	55.8

5-4 全国主要海洋产业就业人员技术构成(2001年)

Technical Composition of Employees at the End of the Year in Major Marine Industries, 2001

单位：万人　Unit: 10 000 persons

海洋产业 Marine Industry	就业人数 Number of the Employed	技术人员 Technical Personnel	技术人员比重(%) Proportion of Technical Personnel(%)
合计 Total	**719.1**	**87.0**	**12.1**
海洋渔业及相关产业 Marine Fishery and the Related Industries	348. 3	10. 0	2. 9
海洋石油和天然气 Offshore Oil and Natural Gas	12. 4	2. 0	16. 1
海滨砂矿 Beach Placer	1. 0	…	1. 7
海洋盐业 Sea Salt Industry	15. 0	0. 6	4. 0
海洋化工 Marine Chemical	16. 1	2. 1	13. 0
海洋生物医药 Marine Biological Pharmaceatical	0. 6	0. 1	16. 7
海洋电力和海水利用 Marine Electric Power and Seawater Utilization	0. 7	0. 1	14. 3
海洋船舶工业 Marine Shipbuilding Industry	20. 6	1. 7	8. 3
海洋工程建筑 Marine Engineering Architecture	38. 8	9. 9	25. 5
海洋交通运输 Maritime Transportation	50. 8	8. 4	16. 5
滨海旅游 Coastal Tourism	78. 3	4. 7	6. 0
其他海洋产业 Other Marine Industries	136. 5	47. 4	34. 7

5-5 全国主要海洋产业就业人员受教育情况(2001年)

Educaton Received by the Employed in the Major Marine Industries, 2001

单位：人 Unit:person

海洋产业 Marine Industry	中专以下 Below the Second Specialized School Level	大学 University	研究生及以上 Postgraduate or above
合计 **Total**	**6 205 000**	**915 000**	**41 000**
海洋渔业及相关产业 Marine Fishery and the Related Industries	3 421 072	29 795	215
海洋石油和天然气 Offshore Oil and Natural Gas	109 631	14 374	141
海滨砂矿 Beach Placer	9 676	38	0
海洋盐业 Sea Salt Industry	143 635	5 618	52
海洋化工 Marine Chemical	144 716	16 443	193
海洋生物医药 Marine Biological Pharmaceatical	3 986	1 700	206
海洋电力和海水利用 Marine Electric Power and Seawater Utilization	5 786	1 198	16
海洋船舶工业 Marine Shipbuilding Industry	188 397	17 109	218
海洋工程建筑 Marine Engineering Architecture	216 319	171 230	347
海洋交通运输 Maritime Transportation	412 369	93 976	1 513
滨海旅游 Coastal Tourism	680 204	100 873	1 928
其他海洋产业 Other Marine Industries	869 209	492 646	36 171

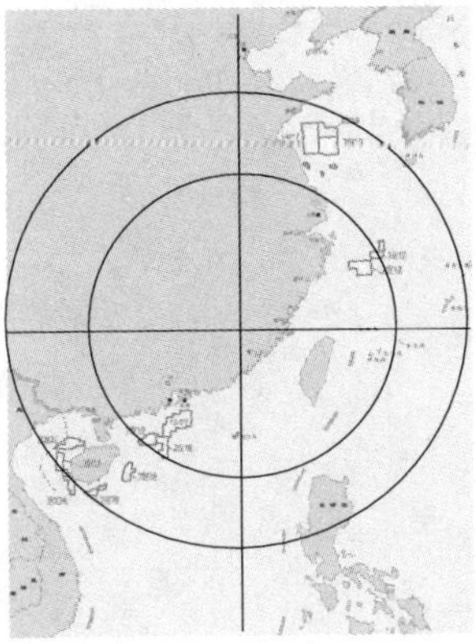

6

海洋科学技术

Marine Science and Technology

6-1 分行业海洋科研机构及人员情况（2006年）
Marine Scientific Research Institutions and Personnel by Industry, 2006

行　业 Industry	机构数（个） Number of Institutions	从业人员（人） Employed Population(person)
合 计 **Total**	**136**	**18 271**
海洋基础科学研究 Marine Basic Science Research	94	13 638
海洋自然科学 Marine Natural Science	54	10 146
海洋社会科学 Marine Social Science	4	631
海洋农业科学 Marine Agricaltural Science	33	2 635
海洋生物医药 Marine Biopharmaceutics	3	226
海洋工程技术研究 Marine Engineering Technology Research	34	4 085
海洋化学工程技术 Marine Chemical Engineeing Technology	7	750
海洋交通运输工程技术 Marine Communications and Transport Technology	10	1 540
海洋能源开发技术 Marine Energies Development Technology	1	377
海洋环境工程技术 Marine Environmental Engineering Technology	11	908
河口水利工程技术 Eustuarine Water Conservancy Projects Technology	3	340
其他海洋工程技术 Other Marine Engineering Technology	2	170
海洋信息服务业 Marine Information Service	8	548

注：机构为县级以上科研机构；行业分类参照《海洋及相关产业分类》标准。

Note:The institutions are the scientific research institutions above the county level;The classification of industries follows the Standard "Classification of Marine Industries and the Related Industries".

6-2 分行业海洋科研机构从事科技活动人员学历构成（2006年）
Educational Background Composition of the Personnel Engaged in Scientific and Technological Activities in Marine Scientific Research Institutions by Industry, 2006

单位：人 Unit:person

行 业 Industry	从事科技活动人员 Personnel Engaged in Scientifical Activities	研究生 Postgraduate	#博士 Doctor	大学生 Graduate	大专生 Junior College for Professional Training	其他 Others
合 计 Total	**13 941**	**4 462**	**1 723**	**5 737**	**2 260**	**1 482**
海洋基础科学研究 Marine Basic Science Research	10 488	3 736	1 652	3 926	1 687	1 139
海洋自然科学 Marine Natural Science	7 805	3 228	1 554	2 697	1 104	776
海洋社会科学 Marine Social Science	553	148	21	256	112	37
海洋农业科学 Marine Agricaltural Science	1 922	305	73	888	428	301
海洋生物医药 Marine Biopharmaceutics	208	55	4	85	43	25
海洋工程技术研究 Marine Engineering Technology Research	3 029	652	65	1 605	473	299
海洋化学工程技术 Marine Chemical Engineeing Technology	432	92	7	226	66	48
海洋交通运输工程技术 Marine Communications and Transport Technology	1 168	236	23	680	169	83
海洋能源开发技术 Marine Energies Development Technology	245	34	2	132	27	52
海洋环境工程技术 Marine Environmental Engineering Technology	797	190	16	421	129	57
河口水利工程技术 Eustuarine Water Conservancy Projects Technology	234	60	13	73	55	46
其他海洋工程技术 Other Marine Engineering Technology	153	40	4	73	27	13
海洋信息服务业 Marine Information Service	424	74	6	206	100	44

6-3 分行业海洋科研机构科技活动人员情况（2006年）
Personnel Engaged in Scientifics and Technological Activities in the Marine Scientific Research Institutions by Industry, 2006

单位：人 Unit:person

行 业 Industry	从事科技活动人员 Personnel Engaged in Scientifical Activities	高级职称 Senior	中级职称 Intermediate	初级职称 Primary	其 他 Others
合 计 **Total**	**13 941**	**5 484**	**4 424**	**2 817**	**1 216**
海洋基础科学研究 Marine Basic Science Research	10 488	4 312	3 342	1 980	854
海洋自然科学 Marine Natural Science	7 805	3 492	2 440	1 282	591
海洋社会科学 Marine Social Science	553	171	232	110	40
海洋农业科学 Marine Agricaltural Science	1 922	605	600	505	212
海洋生物医药 Marine Biopharmaceutics	208	44	70	83	11
海洋工程技术研究 Marine Engineering Technology Research	3 029	1 069	905	709	346
海洋化学工程技术 Marine Chemical Engineering Technology	432	163	107	109	53
海洋交通运输工程技术 Marine Communications and Transport Technology	1 168	471	329	274	94
海洋能源开发技术 Marine Energies Development Technology	245	71	81	40	53
海洋环境工程技术 Marine Environmental Engineering Technology	797	210	298	224	65
河口水利工程技术 Eustuarine Water Conservancy Projects Technology	234	78	63	34	59
其他海洋工程技术 Other Marine Engineering Technology	153	76	27	28	22
海洋信息服务业 Marine Information Service	424	103	177	128	16

6-4 分行业海洋科研机构经常费收入（2006年）
Routine Funds Receipts of the Marine Scientific Research Institutions by Industry, 2006

单位：千元 unit：1 000 yuan

行 业 Industry	收入总额 Total	科技经费筹集额 Amount of Funds Raised for Science and Technology	#政府资金 Funds from Government	生产经营活动收入 Earnings from Production and Business Activities	其他收入 Other Incomes	非科技活动借贷款 Loan for Non-Scientific and Technological Activities
合 计 Total	**5 289 252**	**4 487 636**	**3 105 193**	**420 838**	**374 668**	**6 110**
海洋基础科学研究 Marine Basic Science Research	3 991 362	3 606 913	2 810 721	85 024	299 425	0
海洋自然科学 Marine Natural Science	3 486 541	3 205 177	2 441 169	62 000	219 364	0
海洋社会科学 Marine Social Science	58 838	51 635	49 691	390	6 813	0
海洋农业科学 Marine Agricaltural Science	369 723	322 032	292 930	22 634	25 057	0
海洋生物医药 Marine Biopharmaceutics	76 260	28 069	26 931	0	48 191	0
海洋工程技术研究 Marine Engineering Technology Research	1 216 271	808 741	228 551	331 737	69 793	6 000
海洋化学工程技术 Marine Chemical Engineeing Technology	252 690	94 751	21 178	143 649	8 290	6 000
海洋交通运输工程技术 Marine Communications and Transport Technology	466 338	378 290	66 804	43 685	44 363	0
海洋能源开发技术 Marine Energies Development Technology	143 062	8 471	0	125 906	8 685	0
海洋环境工程技术 Marine Environmental Engineering Technology	238 490	215 703	125 770	15 319	7 468	0
河口水利工程技术 Eustuarine Water Conservancy Projects Technology	70 237	69 466	14 600	0	771	0
其他海洋工程技术 Other Marine Engineering Technology	45 454	42 060	199	3 178	216	0
海洋信息服务业 Marine Information Service	81 619	71 982	65 921	4 077	5 450	110

6-5 分行业海洋科研机构科技课题情况（2006年）

Marine Science and Technology Research Projects of the Research Institutions by Industry, 2006

单位：项　　Unit: item

行　业 Industry	合　计 Total	基础研究 Basic Research	应用研究 Applied Research	试验发展 Experimental Development	成果应用 Result Application	科技服务 Scientific and Technological Service
合 计 **Total**	**6 593**	**1 687**	**1 734**	**1 238**	**699**	**1 235**
海洋基础科学研究 Marine Basic Science	5 810	1 686	1 721	1 079	534	790
海洋自然科学 Marine Natural Science	5 025	1 663	1 622	800	356	584
海洋社会科学 Marine Social Science	29	0	5	7	0	17
海洋农业科学 Marine Agricaltural	724	23	93	270	155	183
海洋生物医药 Marine Biopharmaceutics	32	0	1	2	23	6
海洋工程技术研究 Marine Engineering Technology Research	759	1	13	158	165	422
海洋化学工程技术 Marine Chemical Engineeing Technology	61	0	8	39	12	2
海洋交通运输工程技术 Marine Communications and Transport Technology	431	1	2	67	126	235
海洋能源开发技术 Marine Energies Development Technology	20	0	0	20	0	0
海洋环境工程技术 Marine Environmental Engineering Technology	199	0	3	25	20	151
河口水利工程技术 Eustuarine Water Conservancy Projects Technology	28	0	0	7	7	14
其他海洋工程技术 Other Marine Engineering Technology	20	0	0	0	0	20
海洋信息服务业 Marine Information Service	24	0	0	1	0	23

6-6 分行业海洋科研机构科技论著情况（2006年）

Marine Scientific and Technological Works of the Research Institutions by Industry, 2006

行 业 Industry	发表科技论文（篇） Scientific Theses Published	#国外发表 Published Abroad	出版科技著作（种） Scientific and Technological Works Published (kinds)
合 计 **Total**	**8 492**	**2 048**	**110**
海洋基础科学研究 Marine Basic Science Research	7 806	2 028	100
海洋自然科学 Marine Natural Science	6 454	1 934	72
海洋社会科学 Marine Social Science	353	25	0
海洋农业科学 Marine Agricaltural Science	960	66	28
海洋生物医药 Marine Biopharmaceutics	39	3	0
海洋工程技术研究 Marine Engineering Technology Research	638	13	7
海洋化学工程技术 Marine Chemical Engineeing Technology	117	4	2
海洋交通运输工程技术 Marine Communications and Transport Technology	250	3	3
海洋能源开发技术 Marine Energies Development Technology	32	1	1
海洋环境工程技术 Marine Environmental Engineering Technology	137	5	1
河口水利工程技术 Eustuarine Water Conservancy Projects Technology	102	0	0
其他海洋工程技术 Other Marine Engineering Technology	0	0	0
海洋信息服务业 Marine Information Service	48	7	3

6-7 分行业科研机构科技专利情况（2006年）
Marine Scientific and Technological Patents of the Research Institutions by Industry, 2006

单位：件 Unit: number

行业 Industry	专利申请受理数 Number of Patent Applications Accepted	#发明专利 Patents for Discoveries	专利授权数 Number of Patents Granted	#发明专利 Patents for Discoveries	拥有发明专利总数 Total Number of Patents for Discoveries
合计 Total	**567**	**434**	**379**	**226**	**1 229**
海洋基础科学研究 Marine Basic Science Research	504	393	340	210	1 122
海洋自然科学 Marine Natural Science	446	357	289	181	1 026
海洋社会科学 Marine Social Science	0	0	0	0	0
海洋农业科学 Marine Agricaltural Science	58	36	51	29	96
海洋生物医药 Marine Biopharmaceutics	0	0	0	0	0
海洋工程技术研究 Marine Engineering Technology Research	63	41	39	16	107
海洋化学工程技术 Marine Chemical Engineeing Technology	20	19	9	9	49
海洋交通运输工程技术 Marine Communications and Transport Technology	10	5	6	0	32
海洋能源开发技术 Marine Energies Development Technology	22	13	14	2	5
海洋环境工程技术 Marine Environmental Engineering Technology	11	4	10	5	21
河口水利工程技术 Eustuarine Water Conservancy Projects Technology	0	0	0	0	0
其他海洋工程技术 Other Marine Engineering Technology	0	0	0	0	0
海洋信息服务业 Marine Information Service	0	0	0	0	0

6-8 分地区海洋科研机构及人员情况（2006年）

Marine Scientific Research Institutions and Personnel by Regions, 2006

地　区 Region	机构数（个） Number of Institutions	从业人员（人） Employed Population(person)
合 计 **Total**	**136**	**18 271**
北 京 Beijing	11	3 127
天 津 Tianjin	11	2 656
河 北 Hebei	4	416
辽 宁 Liaoning	8	592
上 海 Shanghai	13	2 500
江 苏 Jiangsu	7	1 213
浙 江 Zhejiang	17	975
福 建 Fujian	10	607
山 东 Shandong	20	3 001
广 东 Guangdong	24	2 254
广 西 Guangxi	6	165
海 南 Hainan	3	78
其 他 Other	2	687

6-9 分地区海洋科研机构及人员学历构成（2006年）
Educational Background Composition of the Personnel Engaged in Scientific and Technological Activities in Marine Scientific Research Institutions by Regions, 2006

单位：人 Unit:person

地 区 Region	从事科技活动人员 Personnel Engaged in Scientifical Activities	研究生 Postgraduate	#博士 Doctor	大学生 Graduate	大专生 Junior College for Professional Training	其他 Others
合 计 Total	**13 941**	**4 462**	**1 723**	**5 737**	**2 260**	**1 482**
北 京 Beijing	2 460	1 259	768	746	334	121
天 津 Tianjin	1 711	309	34	935	255	212
河 北 Hebei	370	79	10	200	62	29
辽 宁 Liaoning	493	88	26	276	88	41
上 海 Shanghai	1 983	592	158	790	377	224
江 苏 Jiangsu	933	331	102	375	132	95
浙 江 Zhejiang	805	179	33	422	119	85
福 建 Fujian	567	136	28	241	78	112
山 东 Shandong	2 200	620	255	843	445	292
广 东 Guangdong	1 761	616	196	648	268	229
广 西 Guangxi	120	4	0	74	29	13
海 南 Hainan	69	8	0	38	13	10
其 他 Other	469	241	113	149	60	19

6-10 分地区海洋科研机构专业技术人员情况（2006年）
Marine Scientific Research Institutions and Personnel by Regions, 2006

单位：人 Unit:person

地 区 Region	专业技术人员 Professional and Technological Personnel	高级职称 Senior	中级职称 Intermediate	初级职称 Primary	其他 Others
合计 Total	**13 941**	**5 484**	**4 424**	**2 817**	**1 216**
北京 Beijing	2 460	1 271	731	354	104
天津 Tianjin	1 711	584	519	450	158
河北 Hebei	370	152	158	51	9
辽宁 Liaoning	493	168	158	116	51
上海 Shanghai	1 983	672	587	443	281
江苏 Jiangsu	933	468	307	118	40
浙江 Zhejiang	805	297	256	193	59
福建 Fujian	567	197	180	138	52
山东 Shandong	2 200	823	702	476	199
广东 Guangdong	1 761	625	595	338	203
广西 Guangxi	120	17	49	36	18
海南 Hainan	69	8	21	32	8
其他 Other	469	202	161	72	34

6-11 分地区海洋科研机构经常费收入（2006年）
Routine Fund Receipts of Marine Scientific Research Institutions by Regions,2006

单位：千元 Unit：1 000 yuan

地　区 Region	收入总额 Total	科技经费筹集额 Amount of Funds Raised for Science and Technology	#政府资金 Funds from Government	生产经营活动收入 Earnings from Production and Business Activities	其他收入 Other Incomes	非科技活动借贷款 Loan for Non-Scientific and Technological Activities
合 计 Total	**5 289 252**	**4 487 636**	**3 105 193**	**420 838**	**374 668**	**6 110**
北 京 Beijing	1 186 183	1 073 166	882 410	30 854	82 163	0
天 津 Tianjin	661 124	420 803	205 630	206 432	33 889	0
河 北 Hebei	39 518	33 085	32 135	72	6 361	0
辽 宁 Liaoning	117 170	99 553	67 836	1 244	16 373	0
上 海 Shanghai	762 216	641 067	397 577	39 332	81 817	0
江 苏 Jiangsu	344 950	330 720	167 297	570	13 660	0
浙 江 Zhejiang	414 235	384 523	279 928	17 571	12 031	110
福 建 Fujian	224 202	215 565	171 338	484	8 153	0
山 东 Shandong	783 307	610 034	432 893	96 590	70 683	6 000
广 东 Guangdong	585 344	524 551	365 122	21 777	39 016	0
广 西 Guangxi	12 729	9 945	7 379	1 883	901	0
海 南 Hainan	10 544	10 444	9 052	0	100	0
其 他 Other	147 730	134 180	86 596	4 029	9 521	0

6-12 分地区海洋科研机构科技课题情况（2006年）
Marine Science and Technology Research Projects of the Research Institutions by Regions, 2006

单位：项　　Unit: item

地 区 Region	合 计 Total	基础研究 Basic Research	应用研究 Applied Research	试验发展 Experimental Development	成果应用 Result Application	科技服务 Scientific and Technological Service
合 计 Total	**6 593**	**1 687**	**1 734**	**1 238**	**699**	**1 235**
北 京 Beijing	1 705	855	586	86	20	158
天 津 Tianjin	342	0	48	97	77	120
河 北 Hebei	49	0	5	10	8	26
辽 宁 Liaoning	62	0	6	11	19	26
上 海 Shanghai	684	75	218	108	101	182
江 苏 Jiangsu	989	55	272	393	244	25
浙 江 Zhejiang	315	50	65	54	56	90
福 建 Fujian	354	71	62	74	51	96
山 东 Shandong	807	199	221	159	56	172
广 东 Guangdong	985	223	215	184	36	327
广 西 Guangxi	24	0	1	16	3	4
海 南 Hainan	33	0	0	0	28	5
其 他 Other	244	159	35	46	0	4

6-13 分地区海洋科研机构科技论著情况（2006年）

Marine Scientific and Technological Works of the Research Institutions by Regions, 2006

地 区 Region	发表科技论文（篇） Scientific Theses Published	#国外发表 Published abroad	出版科技著作（种） Scientific and Technological Works Published (kind)
合 计 Total	**8 492**	**2 048**	**110**
北 京 Beijing	2 722	1 084	23
天 津 Tianjin	320	17	8
河 北 Hebei	391	26	0
辽 宁 Liaoning	86	8	2
上 海 Shanghai	766	128	6
江 苏 Jiangsu	671	101	14
浙 江 Zhejiang	300	42	10
福 建 Fujian	302	59	3
山 东 Shandong	1 167	213	23
广 东 Guangdong	1 181	222	13
广 西 Guangxi	11	0	0
海 南 Hainan	41	0	1
其 他 Other	534	148	7

6-14 分地区海洋科研机构科技专利情况（2006年）
Marine Scientific and Technological Patents of the Research Institutions by Regions, 2006

单位：件 Unit: number

地区 Region	专利申请受理数 Number of Patent Applications Accepted	#发明专利 Patents for Discoveries	专利授权数 Number of Patents Granted	#发明专利 Patents for Discoveries	拥有发明专利总数 Total Number of Patents for Discoveries
合计 Total	**567**	**434**	**379**	**226**	**1 229**
北京 Beijing	183	158	108	84	477
天津 Tianjin	63	29	40	20	74
河北 Hebei	0	0	0	0	0
辽宁 Liaoning	0	0	1	1	2
上海 Shanghai	88	71	63	30	168
江苏 Jiangsu	11	5	8	5	25
浙江 Zhejiang	20	9	9	6	7
福建 Fujian	4	4	1	1	20
山东 Shandong	109	88	99	49	239
广东 Guangdong	82	67	35	24	190
广西 Guangxi	0	0	0	0	0
海南 Hainan	0	0	0	0	0
其他 Other	7	3	15	6	27

主要统计指标解释

1.海洋科研机构 指有明确的研究方向和任务，有一定水平的学术带头人和一定数量、质量的研究人员，有开展研究工作的基本条件，长期有组织地从事海洋研究与开发活动的机构。

2. 从业人员 指由本机构年末直接组织安排工作并支付工资的各类人员总数。包括固定职工、国家有编制的合同制职工、招聘人员和返聘的离退休人员。不包括离退休人员、停薪留职人员。

3.从事科技活动人员 指从业人员中的科技管理人员、课题活动人员和科技服务人员。

其他收入

4.高级职称 指研究员、副研究员；教授、副教授；高级工程师；高级农艺师；正、副主任医(药、护、技)师；高级实验师；高级统计师；高级经济师；高级会计师；编审(正、副编审)；译审(正、副译审)、高级(主任)记者；正、副研究馆员等。

5.中级职称 指助理研究员；讲师；工程师；农艺师；主治医(药、护、技)师；实验师；统计师；经济师；会计师；编辑；翻译；记者；馆员等。

6.初级职称 指研究实习员；助教；助理工程师、技术员；助理农艺师、农业技术员；医(药、护、技)师、医(药、护、技)士；助理实验师、实验员；助理统计师、统计员；助理经济师；助理会计师、会计员；助理编辑、见习编辑；助理翻译；助理记者；助理馆员、管理员等。

7.经常费 包括暂收(暂付)款，经常费收入中各项皆为毛收入。

8.科技经费筹集额 指从各种渠道筹集到的计划用于本单位科技活动的经费，不论来源渠道如何。

9. 政府资金 指由各级政府部门直接拨款或企事业单位利用政府资金委托本机构从事科学技术活动所获得的收入。

10.生产经营活动收入 指本机构在科研、技术等专业业务活动以外开展非独立核算的经营活动取得的收入，包括产品(商品)销售收入、经营服务收入、工程承包收入、租赁收入和其他经营收入。

11.其他收入 指开展科技活动与生产经营活动以外的各项活动的收入，包括：用于离退休人员的政府拨款。

12 非科技活动借贷款 指本机构为开展非科技活动从各种渠道获得的各类借、贷款。不论偿还形式、期限和数额如何，均按当年获得的借、贷款额填报。不包括基本建设贷款。

13.基础研究 为获得新知识而进行的独创性研究。其目的是揭示观察到的现象和事实的基本原理和规律，而不以任何特定的实际应用为目的。

14.应用研究 为获得新的科学技术知识而进行的独创性研究。它主要针对某一特定的实际应用目的。应用研究通常是为了确定基础研究成果或知识的可能的用途，或是为达到某一具体的、预定的实际目的确定新的方法(原理性)或途径。

15.试验发展 利用从研究或实际经验获得的知识，为生产新的材料、产品和装置，建立新的工艺和系统以及对已生产或建立的上述各项进行实质性的改进而进行的系统性工作。

16.成果应用 为解决 R&D 活动阶段产生的新产品、新装置、新工艺、新技术、新方法、新系统和服务等能投入生产或在实际应用中所存在的技术问题而进行的系统性活动。它不具有创新成分。此类活动包括为达到生产目的而进行的定型设计和试制，以及为扩大新产品的生产规模和探索新方法、新技术、新工艺等的应用领域而进行的适应性试验。
17.科技服务 与科学研究与实验发展有关，并有助于科学技术知识的产生、传播和应用的活动。包括为扩大科技成果的使用范围而进行的示范性推广工作；为用户提供科技信息和文献服务的系统性工作；为用户提供可行性报告、技术方案、建议及进行技术论证等技术咨询工作；自然、生物现象的日常观测、监测，资源的考查和勘探；有关社会、人文、经济现象的通用资料的收集，如统计、市场调查等，以及这些资料的常规分析与整理；为社会和公众提供的测试、标准化、计量、计算、质量控制和专利服务，不包括工商企业为进行正常生产而开展的上述活动。
18.生产性活动 由于业务特殊的工艺设备条件，或掌握某种技术专长或诀窍，所进行的小量非常规生产。
19.科技论文 在全国性学报或学术刊物上、省部属大专院校对外正式发行的学报或学术刊物上发表的论文，以及向国外发表的论文。
20.科技著作 经过正式出版部门编印出版的科技专著、大专院校教科书、科普著作。
21.专利申请受理数 当年本单位向专利管理部门提出申请并被受理的职务专利申请件数。
22.专利授权数 当年由专利管理部门授予本单位专利权的职务专利件数。

Explanatory Notes on Main Statistical Indicators

1.Marine Scientific Research Institution refers to the institution which has definite research orientations and tasks, high-level academic leading personnel and fair-sized, qualified research personnel, and basic conditions for research work and which is engaged for a long time in the marine research and development activities in an organized way.
2.Employees refer to the total number of personnel of various kinds employed and paid by the institution at the end of the year, including fixed employees, contract workers of staff belonging to the state authorized staff , recruited personnel, reemployed retired personnel, but not including the retired and the personnel on leave with pay suspension.
3.Personnel Engaged in Scientific and Technological Activities refers to the personnel for scientific and technological management,personnel engaged in the activities of research topics and scientific and technological service personnel.
4.Senior Technical Title refers to research scientist, associate research scientist; professor, associate professor; senior engineer; senior agronomist; professor-rank and associate professor-rank doctor (pharmacists, nurses and technicians); senior laboratory technician; senior

statisticians; senior economic engineer; chief accountant; senior editor (professor and associate professor ranks); senior translator (professor and associate professor ranks); senior journalist; research librarian (professor and associate professor ranks), etc.

5.Intermediate Technical Title refers to assistant research scientist; lecturer; engineer; agronomist; lecturer-rank doctor (pharmacist, nurse and technician); laboratory technician; lecturer-rank statistician; economic engineer; accountant; editor; translator; journalist; librarian, etc.

6.Primary Technical Title refers to trainee researcher; assistant; assistant engineer, technician, assistant agronomist, agricultural technician; assistant-rank doctor (pharmacist, nurse and technician); assistant laboratory technician; assistant statistician; assistant economic engineer, assistant accountant; assistant editor, editor on probation; assistant translator; assistant journalist; assistant research librarian, librarian ,etc.

7.Routine Funds include the money temporarily collected (paid),and the various items of routine funds receipts are all gross income.

8.Amount of Scientific and Technological Funds Raised refers to the funds raised through all channels planned to be used as funds for the scientific and technological activities in the institution regardless of their source and channels.

9.Funds from Government refers to the direct appropriations by the government departments at all levels or the earnings from conducting scientific and technological activities entrusted to the institution by enterprises and institutions by earning the funds from government.

10.Earnings from Production and Business Activities refer to the incomes obtained from the non-independent accounting business activities carried out by the institution beyond the scientific research,technological and professional activities,including these from sale of products (goods), business and service, contracted projects,leasing and other business.

11.Other Incomes refers to those from the various activities conducted other than scientific and technological activities and production and business activities,including the government appropriations for retired personnels.

12.Loan for Non–scientific and Technological Activities refers to the various types loan obtained by the institutions through all channels for carrying out non-scientific and technological activities,not including the loan for capital construction.The loan, irrespective of its form of reimbursement,term and amount is filled in a form and submitted to the authorities as the amount acquired in the current year.

13.Basic Research refers to the original research to acquire new knowledge. It is aimed at revealing the basic principles and laws of the phenomena and facts observed, but not at any specific practical applications.

14.Applied Research refers to the original research to acquire new scientific and technological knowledge. It mainly serves the purpose of a particular practical application. The purpose of applied research is usually to define the potential uses of the research finds or knowledge obtained from basic research or to identify new methods (principles) or ways to reach a specific and predetermined goal.

15.Experimental Development refers to the systematic work carried out to establish new technologies and systems for producing new materials, products and equipment by using the knowledge obtained from research or practical experience, or to make substantial improvement of the above–mentioned which have been produced or established.

16.Result Application refers to the systematic activities carried out to solve the technical problems that might crop up in the production or practical application of the new products, devices, technologies, techniques, methods, systems and service occurring in the course of R&D activities. They do not bring forth new ideas. Such activities include the finalizing design and trial-production for the purpose of production as well as the adaptive tests to expand the production scale of new products and the application areas of new methods, techniques and technologies.

17.Scientific and Technological Service refers to the activities that are associated with the scientific research and experimental development, and contribute to the generation, dissemination and application of scientific and technological knowledge, which include the demonstrative work of popularization to enlarge the use scope of scientific and technological achievements; the systematic work of providing users with scientific and technological information and literature service; the technical consultation work of providing users with feasibility reports, technical schemes and recommendations and carrying out technical demonstration; routine observation and monitoring of natural and biological phenomena, and the survey and exploration of resources; collection of universal data on the appropriate social, cultural and economic phenomena, such as statistics and market survey, as well as routine analysis and sorting-out of these data; the provision for the society and the public of such service as testing, standardization, computation, quality control and patent, but not including the type of the above-mentioned activities carried out by industrial and commercial enterprises for the purpose of normal production.

18.Productive Activity refers to the small-scale and non-conventional productions due to the presence of special technologies and equipment or mastery of a particular technical expertise or secret of success.

19. Scientific Treatises refer to the theses published in the national journals or academic publications, those officially issued journals or academic publications by universities and colleges under provinces or ministries as well as theses published abroad.

20.Scientific and Technological Works refer to the scientific and technological monographs, textbooks for universities and colleges and popular science books published by the official publishing houses.

21. Number of Patents Applied and Accepted refers to the number of professional patent applications of the unit to the patent administrative department and accepted by it in the year.

22. Number of Patents Granted refers to the number of the professional patents granted to the unit by the patent administrative department in the year.

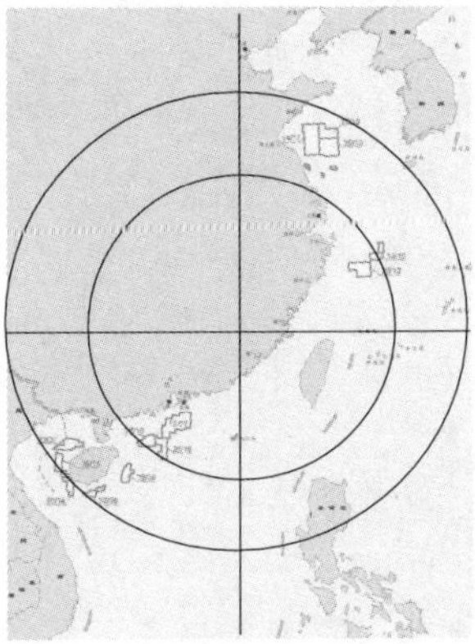

7

海 洋 教 育

Marine Education

7-1 全国各海洋专业博士研究生情况（2006年）

Doctors in Institutions of Higher Education by Oceanography Specialities, 2006

专 业 Speciality	专业点数（个） Number of Speciality Agencies	学生数（人） Students (person)			
		毕业生 Graduated	招 生 Entrants	在校生 Enrollment	毕业班学生 Graduation
合 计 **Total**	**58**	**330**	**465**	**1 841**	**789**
物理海洋学 Physical Oceanography	4	53	40	166	81
海洋化学 Marine Chemistry	4	20	26	121	63
海洋生物学 Marine Biology	6	122	105	378	170
海洋地质 Marine Geology	8	41	60	205	100
海洋科学类新专业 New Speciality of Marine Science	4	1	41	118	31
港口海岸及近海工程 Coastal Harbour and Offshore Engineering	7	17	51	228	71
船舶与海洋结构物设计制造 Ship and Marine Structures Design	10	42	76	333	165
轮机工程 Turbine Engineering	8	13	33	182	61
水声工程 Marine Acoustical Engineering	6	17	31	106	46
船舶与海洋工程新专业 New Speciality of Ship and Marine Engineering	1	4	2	4	1

7-2 全国各海洋专业硕士研究生情况（2006年）

Masters in Institutions of Higher Education by Oceanography Specialities, 2006

专 业 Speciality	专业点数（个） Number of Speciality Agencies	学生数（人）Students (person)			
		毕业生 Graduated	招 生 Entrants	在校生 Enrollment	毕业班学生 Graduation
合 计 **Total**	**109**	**940**	**1 603**	**4 234**	**1 235**
物理海洋学 Physical Oceanography	9	74	103	292	97
海洋化学 Marine Chemistry	9	68	101	295	94
海洋生物学 Marine Biology	15	90	214	590	174
海洋地质 Marine Geology	12	61	116	356	105
海洋科学类新专业 New Speciality of Marine Science	3	1	15	27	0
港口海岸及近海工程 Coastal Harbour and Offshore Engineering	17	162	269	668	171
船舶与海洋结构物设计制造 Ship and Marine Structures Design	15	240	364	968	298
轮机工程 Turbine Engineering	14	150	232	595	184
水声工程 Marine Acoustical Engineering	11	93	163	388	97
船舶与海洋工程新专业 New Specialities in Shipping and Marine Engineering	4	1	26	55	15

7-3 全国各海洋专业本、专科学生情况（2006年）

Students in Institutions of Higher Education and Students in Colleges for Professional Training by Oceanography Specialities, 2006

专 业 Speciality	专业点数（个） Number of Speciality Agencies	学生数（人）Students (person)			
		毕业生 Graduated	招 生 Entrants	在校生 Enrollment	毕业班学生 Graduation
合 计 Total	**249**	**13 203**	**20 197**	**63 834**	**15 848**
海洋科学 Oceanography	13	263	505	1 817	337
海洋技术 Marine Technology	11	279	446	1 546	297
海洋管理 Marine Management	2	18	39	239	68
军事海洋学 Military Oceanography	1	18		32	15
海洋生物资源与环境 Marine Living Resources and Environment	1		38	119	
海洋科学类新专业 New Speciality of Marine Science	1	0	104	104	0
港口航道与海岸工程 Harbour Channel and Coastal Engineering	19	815	1 008	3 743	848
港口海岸及治河工程 Coastal Harbour and River Engineering	1	30		86	38
水资源与海洋工程 Water Resource and Ocean Engineering	1	0	32	144	44
航海技术 Nautical Technology	31	3 381	5 293	14 961	3 804
轮机工程 Turbine Engineering	42	3 674	6 564	18 222	4 611
海事管理 Maritime Affairs Manegernent	7	204	389	1416	375
船舶与海洋工程 Ship and Marine Engineering	20	1 104	1 988	6 968	1 422
海洋工程类新专业 New Speciality of Marine Engineering					
水产养殖学 Aquaculture	76	2 654	3 231	11 557	3 063
海洋渔业科学与技术 Science and Technology of Marine Fishery	13	501	396	1 590	502
水族科学与技术 Science and Technology of Aquatic Animals	2	0	164	582	0
水产类新专业 New Speciality in Aquatic Products	8	262	0	708	424

7-4 全国中等职业教育(学校)各海洋专业学生情况（2006年）
Students in Professional Secondary Schools by Oceanography Specialities, 2006

专 业 Speciality	专业点数(个) Number of Speciality Agencies	学生数（人）Students (person)			
		毕业生 Graduated	招 生 Entrants	在校生 Enrollment	毕业班学生 Graduation
合 计 **Total**	**194**	**6 277**	**12 263**	**28 369**	**8 542**
水产养殖 Aquaculture	46	2 115	2 378	4 346	1 775
航海捕捞 Marine Fishing	4	13	83	300	40
水文与水资源 Hydrology and Water Resources	5	121	0	102	82
船体建造与修理 Ship Hull Building and Repair	18	258	1 784	3 620	779
船舶机械装置 Ship Mechanical Equipment	10	130	443	1 276	483
船舶驾驶 Ship Piloting	43	1 588	3 079	7 243	2 173
轮机管理 Engines Management	31	962	1 955	5 352	1 571
船舶水手与机工 Ship Sailors and Mechanics	28	918	2 073	4 787	1 175
外轮理货 Foreign Ships Freight Forwarding	4	109	207	957	348
工程潜水 Engineering Diving	3	31	52	142	81
船舶电子设备 Ship Electronic Equipment	2	32	209	244	35

7-5 全国成人高等教育海洋专业学生情况（2006年）

Students in Continuing Education by Oceanography Specialities, 2006

专 业 Speciality	专业点数(个) Number of Speciality Agencies	学生数（人）Students (person)			
		毕业生 Graduated	招 生 Entrants	在校生 Enrollment	毕业班学生 Graduation
合 计 **Total**	**99**	**745**	**7 444**	**15 142**	**3 277**
海洋科学类新专业 New Speciality of Marine Science	1	0	0	15	13
港口航道与海岸工程 Harbour Channel and Coastal Engineering	5	0	88	282	99
航海技术 Navigation Technology	27	274	4 310	7 716	1 329
轮机工程 Turbine Engineering	22	230	1 766	3 344	609
海事管理 Maritime Affairs Management	6	1	94	458	188
船舶与海洋工程 Ship and Marine Engineering	11	156	631	1 361	394
海洋工程类新专业 New Speciality of Marine Engineering	2	0	10	231	0
水产养殖学 Aquaculture	20	16	479	1 523	516
海洋渔业科学与技术 Science and Technology of Marine Fishery	1	0	0	40	40
水产类新专业 New Speciality in Aquatic Products	4	68	66	172	89

7-6 分地区各海洋专业博士研究生情况（2006年）

Doctors in Institutions of Higher Education by Regions, 2006

地 区 Region	专业点数(个) Number of Speciality Agencies	学生数（人）Students (person)			
		毕业生 Graduated	招 生 Entrants	在校生 Enrollment	毕业班学生 Graduation
合 计 Total	**58**	**330**	**465**	**1 841**	**789**
北 京 Beijing	1	0	3	12	6
天 津 Tianjin	3	3	10	34	12
河 北 Hebei	0	0	0	0	0
辽 宁 Liaoning	4	11	55	254	62
上 海 Shanghai	7	18	30	139	79
江 苏 Jiangsu	5	3	30	123	27
浙 江 Zhejiang	1	0	0	5	2
福 建 Fujian	5	13	17	68	27
山 东 Shandong	11	184	188	716	333
广 东 Guangdong	6	38	54	177	74
其 他 Others	15	60	78	313	167

7-7 分地区各海洋专业硕士研究生情况（2006年）

Masters in Institutions of Higher Education by Regions, 2006

地　区 Region	专业点数(个) Number of Speciality Agencies	学生数（人）Students (person)			
		毕业生 Graduated	招 生 Entrants	在校生 Enrollment	毕业班学生 Graduation
合 计 Total	**109**	**940**	**1 603**	**4 234**	**1 235**
北 京 Beijing	6	22	34	95	25
天 津 Tianjin	2	51	59	183	61
河 北 Hebei	0	0	0	0	0
辽 宁 Liaoning	9	128	203	550	149
上 海 Shanghai	17	97	153	385	117
江 苏 Jiangsu	7	69	141	366	94
浙 江 Zhejiang	11	35	84	205	56
福 建 Fujian	7	21	66	197	64
山 东 Shandong	15	182	274	778	230
广 东 Guangdong	9	33	65	191	63
其 他 Others	26	302	524	1 284	376

7-8 分地区各海洋专业本、专科学生情况（2006年）

Students in Institutions of Higher Education and Students in Colleges for Professional Training by Regions, 2006

地 区 Region	专业点数(个) Number of Speciality Agencies	学生数（人）Students (person)			
		毕业生 Graduated	招 生 Entrants	在校生 Enrollment	毕业班学生 Graduation
合 计 Total	**249**	**13 203**	**20 197**	**63 834**	**15 848**
北 京 Beijing	1		59	207	30
天 津 Tianjin	12	522	1 043	2 828	591
河 北 Hebei	12	379	479	1 515	497
辽 宁 Liaoning	18	1 486	1 932	6 807	1 645
上 海 Shanghai	12	852	1 087	3 910	918
江 苏 Jiangsu	23	1 876	3 273	9 093	2 373
浙 江 Zhejiang	23	489	1 213	3 469	733
福 建 Fujian	17	1 076	1 776	5 687	1 438
山 东 Shandong	33	1 264	2 506	7 530	1 771
广 东 Guangdong	19	1 342	915	4 066	1 366
广 西 Guangxi	2	51	72	207	37
海 南 Hainan	1	53	60	233	65
其 他 Others	76	3 813	5 782	18 282	4 384

7-9 分地区中等职业教育(学校)各海洋专业学生情况（2006年）

Students in Professional Secondary Schools by Regions, 2006

地 区 Region	专业点数(个) Number of Speciality Agencies	学生数（人）Students (person)			
		毕业生 Graduated	招 生 Entrants	在校生 Enrollment	毕业班学生 Graduation
合 计 Total	**194**	**6 277**	**12 263**	**28 369**	**8 542**
北 京 Beijing	1	19	0	20	11
天 津 Tianjin	4	0	281	385	100
河 北 Hebei	6	124	350	887	261
辽 宁 Liaoning	14	295	567	1 529	634
上 海 Shanghai	7	189	388	1 478	486
江 苏 Jiangsu	25	573	1 579	4 324	989
浙 江 Zhejiang	16	793	1 224	3 659	1 303
福 建 Fujian	27	622	1 786	4 357	1 130
山 东 Shandong	17	724	785	2 177	881
广 东 Guangdong	11	200	583	1 326	195
广 西 Guangxi	7	381	305	801	194
海 南 Hainan	3	98	41	94	32
其 他 Others	56	2 259	4 374	7 332	2 326

7-10 分地区成人高等教育海洋专业学生情况（2006年）
Students in Continuing Education by Regions, 2006

地 区 Region	专业点数(个) Number of Speciality Agencies	学生数（人）Students (person)			
		毕业生 Graduated	招 生 Entrants	在校生 Enrollment	毕业班学生 Graduation
合 计 Total	**99**	**745**	**7 444**	**15 142**	**3 277**
北 京 Beijing	0	0	0	0	0
天 津 Tianjin	1	0	183	372	47
河 北 Hebei	0	0	0	0	0
辽 宁 Liaoning	17	330	1 756	3 607	809
上 海 Shanghai	10	33	604	1 906	720
江 苏 Jiangsu	14	159	1 047	2 179	425
浙 江 Zhejiang	9	4	2 528	3 405	219
福 建 Fujian	8	106	344	777	293
山 东 Shandong	8	34	312	1 163	349
广 东 Guangdong	10	0	260	634	204
广 西 Guangxi	1	0	0	8	8
海 南 Hainan					
其 他 Others	21	79	410	1 091	203

7-11 分地区开设海洋专业高等学校教职工数（2006年）
Number of Teaching and Administrative Staff in the Universities and Colleges Offering Marine Specialities by Regions, 2006

地　区 Region	机构数（个） Number of Institutions	教职工数 Employed Population(person)	专任教师数 Professional (person)
合 计 **Total**	**125**	**243 107**	**127 575**
北 京 Beijing	1	1 225	728
天 津 Tianjin	6	10 350	5 215
河 北 Hebei	6	10 527	5 799
辽 宁 Liaoning	6	8 750	4 641
上 海 Shanghai	5	17 991	7 345
江 苏 Jiangsu	14	29 551	15 306
浙 江 Zhejiang	8	16 562	8 027
福 建 Fujian	5	6 519	4 211
山 东 Shandong	12	19 672	11 372
广 东 Guangdong	8	13 463	7 872
广 西 Guangxi	2	4 369	2 268
海 南 Hainan	1	1 366	820
其 他 Others	51	102 762	53 971

主要统计指标解释

海洋专业 指高等学校和中等职业教育(学校)所设的与海洋有关的专业。

Explanatory Notes on Main Statistical Indicators

Marine Speciality refers to the ocean-related speciality in the institutions of high learning and the professional secondary schools.

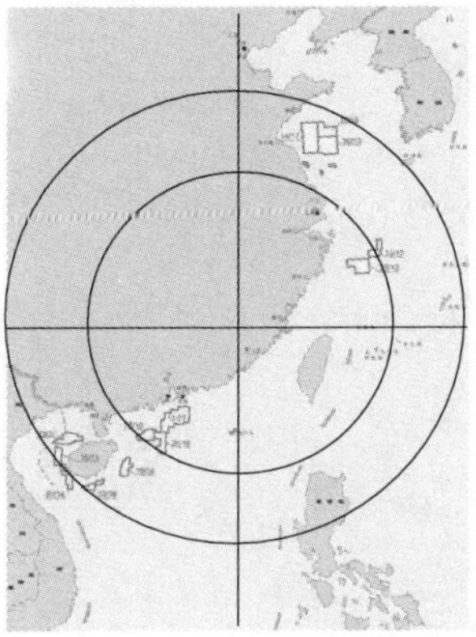

8 海洋环境保护

Marine Environmental Protection

8-1 海区海水质量监测结果（2006年）
Seawater Quality Results of Survey, 2006

单位：毫克/升，叶绿素a为微克/升
Unity: mg/L Chl a: μ g/L

项 目 Item	渤 海 Bohai Sea	黄 海 Huanghai Sea	东 海 East China Sea	南 海 South China Sea
盐 度 Salinity	30. 04	29. 98	28. 67	28. 60
pH	8. 08	8. 17	8. 18	8. 05
化学需氧量 COD	1. 21	0. 82	1. 04	0. 90
溶解氧 DO	6. 50	7. 11	5. 97	6. 10
石油类 Oils	0. 05	0. 04	0. 05	0. 04
活性磷酸盐 Active Phosphate	0. 03	0. 03	0. 03	0. 01
亚硝酸盐氮 Nitrite-nitrogen	0. 03	0. 02	0. 02	0. 02
硝酸盐氮 Nitrate-nitrogan	0. 17	0. 14	0. 46	0. 11
氨 氮 Nitrogen Amide	0. 06	0. 04	0. 03	0. 04
无机氮 Inorganic Nitrogen	0. 27	0. 19	0. 52	0. 17
叶绿素a Chl a	2. 82	2. 92	3. 59	2. 08

8-2 海区海水水质评价结果（2006年）
Seawater Quality Results of Assessment, 2006

项 目 Item	全国总计 National Total	渤 海 Bohai Sea	黄 海 Huanghai Sea	东 海 East China Sea	南 海 South China Sea
测站个数（个） Number of Surrey Stations	**486**	108	93	127	158
较清洁海域面积（万平方公里） Cleaner Sea Area ($10\ 000km^2$)	**5.10**	0.82	1.73	2.09	0.47
轻度污染海域面积（万平方公里） Lightly-Polluted Sea Area ($10\ 000km^2$)	**5.21**	0.74	1.21	2.31	0.96
中度污染海域面积（万平方公里） Moderately-Polluted Sea Area ($10\ 000km^2$)	**1.74**	0.18	0.48	0.84	0.25
严重污染海域面积（万平方公里） Heavily-Polluted Sea Area ($10\ 000km^2$)	**2.84**	0.28	0.92	1.47	0.17
首要超标污染物 Prime Pollutants Exceeding the Set Standard	**无机氮**	无机氮	无机氮	活性磷酸盐	无机氮

8-3 沿海区域废弃物海洋倾倒情况（2006年）

Ocean Dumping of Wastes by Coastal Regions, 2006

地　区 Region		疏浚物（万立方米）Dredged Material ($10\ 000m^3$)	惰性无机地质废料（立方米）Inert, Inorganic Geologic Wastes (m^3)	人体骨灰（盒）Human Ashes (case)
总　计 Total		**17 073**		
环渤海经济区 Round-the-Bohai Sea Economic Zone	**合　计 Total**	**6 724**		
	辽　宁 Liaoning	1 922		
	河　北 Hebei	3 325		
	天　津 Tianjin	153		
	山　东 Shandong	1 324		
长江三角洲经济区 Yangzi River Delta Economic Zone	**合　计 Total**	**5 949**		
	江　苏 Jiangsu	569		
	上　海 Shanghai	3 776		
	浙　江 Zhejiang	1 604		
海峡西岸经济区 Economic Zone on West Side of the Straits	**合　计 Total**	**363**		
	福　建 Fujian	363		
珠江三角洲经济区 Zhujiang River Delta Economic Zone	**合　计 Total**	**3 828**		
	广　东 Guangdong	3 828		
环北部湾经济区 Round-the-Beibu Gulf Economic Zone	**合　计 Total**	**209**		
	广　西 Guangxi	126		
	海　南 Hainan	83		

8-4 海区海洋石油勘探开发污染物排放入海情况（2006年）
Discharge of Pollutants into the Sea from Offshore Oil Exploration and Exploitation, 2006

项　目 Item	**合　计 Total**	渤　海 Bohai Sea	黄　海 Huanghai Sea	东　海 East China Sea	南　海 South China Sea
生产污水（立方米） Sewage from Production(m^3)	**105 897 512**	10 027 520		1 610 000	94 259 992
泥　浆（立方米） Sludge(m^3)	**78 678**	22 868		1 963	53 847
钻　屑（立方米） Debris from Drilling(m^3)	**47 367**	30 406		1 023	15 938
机舱污水（立方米） Sewage from Engineroom(m^3)	**10 491**	425		179	9 887
食品废弃物（立方米） Food Wastes(m^3)	**773**	116		291	366
生活污水（立方米） Domestic Sewage(m^3)	**228 825**	98 041		17 837	112 947
使用消油剂（公斤） Use of Oil Expellant(kg)	**622**	622		0	0

8-5 沿海地区工业废水排放及处理情况（2006年）
Discharge and Treatment of Industrial Waste Water by Coastal Regions, 2006

单位：万吨　Unit:10 000 tons

地　区 Region		工业废水排放总量 Total Volume of Industrial Waste Water Discharged	直接排入海的 Discharged Directly to Sea	工业废水中 Including 符合排放标准的 Up-to-Standard	达标率 (%) Standardized Rate（%）
总 计 Total		**1 426 096**	**131 736**	**1 325 165**	**92.9**
环渤海经济区 Round-the-Bohai Sea Economic Zone	**合 计 Total**	**392 407**	**42 726**	**374 222**	**95.4**
	辽 宁 Liaoning	94 724	32 706	88 007	92.9
	河 北 Hebei	130 340	670	121 750	93.4
	天 津 Tianjin	22 978	678	22 925	99.8
	山 东 Shandong	144 365	8 672	141 540	98.0
长江三角洲经济区 Yangzi River Delta Economic Zone	**合 计 Total**	**535 110**	**23 339**	**500 017**	**93.4**
	江 苏 Jiangsu	287 181	695	280 457	97.7
	上 海 Shanghai	48 336	13 508	47 146	97.5
	浙 江 Zhejiang	199 593	9 136	172 414	86.4
海峡西岸经济区 Economic Zone on West Side of the Straits	**合 计 Total**	**127 583**	**56 615**	**124 960**	**97.9**
	福 建 Fujian	127 583	56 615	124 960	97.9
珠江三角洲经济区 Zhujiang River Delta Economic Zone	**合 计 Total**	**234 713**	**4 216**	**199 215**	**84.9**
	广 东 Guangdong	234 713	4 216	199 215	84.9
环北部湾经济区 Round-the-Beibu Gulf Economic Zone	**合 计 Total**	**136 283**	**4 840**	**126 751**	**93.0**
	广 西 Guangxi	128 932	744	119 795	92.9
	海 南 Hainan	7 351	4 096	6 956	94.6

8-6 沿海城市工业废水排放及处理情况（2006年）
Discharge and Treatment of Industrial Waste Water by Coastal Cities, 2006

单位：万吨 Unit:10 000 tons

沿海城市 Coastal City	工业废水排放总量 Total Volume of Industrial Waste Water Discharged	工业废水中 Including		
		直接排入海的 Discharged Directly to Sea	符合排放标准的 Up-to-Standard	达标率（%） Standardized Rate（%）
天 津 Tianjin	22 978	678	22 925	99.8
唐 山 Tangshan	28 713	240	27 680	96.4
秦皇岛 Qinhuangdao	5 508	222	5 296	96.2
沧 州 Cangzhou	7 469	0	7 221	96.7
大 连 Dalian	33 698	30 798	32 986	97.9
丹 东 Dandong	3 783	172	3 147	83.2
锦 州 Jinzhou	5 087	93	4 231	83.2
营 口 Yingkou	5 330	765	4 987	93.6
盘 锦 Panjin	1 796	0	1 655	92.1
葫芦岛 Huludao	3 134	734	3 109	99.2
上 海 Shanghai	48 336	13 508	47 146	97.5
南 通 Nantong	15 376	0	15 376	100.0
连云港 Lianyungang	3 855	637	3 779	98.0
盐 城 Yancheng	9 124	3	8 961	98.2

续表 continued

沿海城市 Coastal City	工业废水排放总量 Total Volume of Industrial Waste Water Discharged	工业废水中 Including		
		直接排入海的 Discharged Directly to Sea	符合排放标准的 Up-to-Standard	达标率（%） Standardized Rate（%）
杭　州　Hangzhou	76 539	0	55 811	72.9
宁　波　Ningbo	15 827	5 565	14 506	91.7
温　州　Wenzhou	11 409	584	10 181	89.2
嘉　兴　Jiaxing	16 291	240	15 713	96.5
绍　兴　Shaoxing	27 867	159	27 334	98.1
舟　山　Zhoushan	1 754	1 641	1 595	90.9
台　州　Taizhou	4 892	946	4 424	90.4
福　州　Fuzhou	5 631	280	5 418	96.2
厦　门　Xiamen	3 979	241	3 862	97.1
莆　田　Putian	1 578	479	1 485	94.1
泉　州　Quanzhou	26 894	889	26 843	99.8
漳　州　Zhangzhou	58 118	53 594	57 793	99.4
宁　德　Ningde	2 991	1 132	2 406	80.4
青　岛　Qingdao	9 521	2 511	9 401	98.7
东　营　Dongying	8 953	0	8 863	99.0
烟　台　Yantai	7 066	2 552	6 896	97.6
潍　坊　Weifang	12 744	0	12 595	98.8
威　海　Weihai	2 851	1 075	2 823	99.0
日　照　Rizhao	6 764	2 533	6 764	100.0
滨　州　Binzhou	10 401	0	9 797	94.2

续表 continued

沿海城市 Coastal City	工业废水排放总量 Total Volume of Industrial Waste Water Discharged		工业废水中 Including	
		直接排入海的 Discharged Directly to Sea	符合排放标准的 Up-to-Standard	达标率（%） Standardized Rate（%）
广　州 Guangzhou	20 445	0	19 629	96. 0
深　圳 Shenzhen	6 396	844	6 156	96. 2
珠　海 Zhuhai	3 434	328	3 301	96. 1
汕　头 Shantou	5 437	346	5 218	96. 0
江　门 Jiangmen	12 127	241	10 942	90. 2
湛　江 Zhanjiang	6 620	300	4 848	73. 2
茂　名 Maoming	7 496	1 469	6 614	88. 2
惠　州 Huizhou	6 204	311	5 958	96. 0
汕　尾 Shanwei	3 818	2	2 528	66. 2
阳　江 Yangjiang	1 710	26	1 509	88. 2
东　莞 Dongguan	24 134	0	22 344	92. 6
中　山 Zhongshan	10 284	250	9 842	95. 7
潮　州 Chaozhou	1 749	0	1 463	83. 6
揭　阳 Jieyang	3 048	2	2 727	89. 5
北　海 Beihai	2 653	26	2 379	89. 7
防城港 Fangchenggang	2 227	432	1 875	84. 2
钦　州 Qinzhou	4 291	280	3 977	92. 7
海　口 Haikou	627	0	619	98. 7
三　亚 Sanya	292	0	288	98. 6

8-7 沿海地带工业废水排放及处理情况（2006年）
Discharge and Treatment of Industrial Waste Water by Coastal Counties, 2006

单位：万吨　Unit:10 000 tons

沿海地带 Coastal County（District）	工业废水排放总量 Total Volume of Industrial Waste Water Discharged	直接排入海的 Discharged Directly to Sea	工业废水中 Including 符合排放标准的 Up-to-Standard	达标率（%） Standardized Rate（%）
天　津　Tianjin	**9 515**	**678**	**9 500**	**99.8**
河　北　Hebei	**19 603**	**222**	**19 099**	**97.4**
唐　山　Tangshan	13 950	0	13 685	98.1
秦皇岛　Qinhuangdao	4 428	222	4 254	96.1
沧　州　Cangzhou	1 225	0	1 161	94.8
辽　宁　Liaoning	**37 393**	**27 681**	**35 776**	**95.7**
大　连　Dalian	28 028	26 010	27 431	97.9
丹　东　Dandong	929	172	893	96.1
锦　州　Jinzhou	1 718	0	1 008	58.7
营　口　Yingkou	3 336	765	3 183	95.4
盘　锦　Panjin	390	0	278	71.3
葫芦岛　Huludao	2 991	734	2 983	99.7
上　海　Shanghai	**31 823**	**13 378**	**30 853**	**97.0**
江　苏　Jiangsu	**17 985**	**630**	**16 906**	**94.0**
南　通　Nantong	8 132	0	7 343	90.3
连云港　Lianyungang	3 811	627	3 737	98.1
盐　城　Yancheng	6 041	3	5 826	96.4

续表 continued

沿海地带 Coastal County (District)	工业废水排放总量 Total Volume of Industrial Waste Water Discharged		工业废水中 Including	
		直接排入海的 Discharged Directly to Sea	符合排放标准的 Up-to-Standard	达标率（%） Standardized Rate（%）
浙　江 Zhejiang	**76 875**	**7 202**	**64 541**	**84.0**
杭　州 Hangzhou	23 707	0	14 261	60. 2
宁　波 Ningbo	12 626	3 658	11 804	93. 5
温　州 Wenzhou	9 057	564	7 955	87. 8
嘉　兴 Jiaxing	7 044	240	6 813	96. 7
绍　兴 Shaoxing	19 104	153	18 889	98. 9
舟　山 Zhoushan	1 754	1 641	1 595	90. 9
台　州 Taizhou	3 583	946	3 224	90. 0
福　建 Fujian	**89 339**	**56 526**	**88 568**	**99.1**
福　州 Fuzhou	3 446	267	3 343	97. 0
厦　门 Xiamen	3 979	241	3 862	97. 1
莆　田 Putian	1 578	479	1 485	94. 1
泉　州 Quanzhou	22 993	889	22 942	99. 8
漳　州 Zhangzhou	55 488	53 536	55 384	99. 8
宁　德 Ningde	1 855	1 114	1 553	83. 7
山　东 Shandong	**32 148**	**6 547**	**31 725**	**98.7**
青　岛 Qingdao	3 355	447	3 236	96. 5
东　营 Dongying	8 953	0	8 863	99. 0
烟　台 Yantai	5 993	2 492	5 920	98. 8
潍　坊 Weifang	4 021	0	3 988	99. 2
威　海 Weihai	2 851	1 075	2 823	99. 0
日　照 Rizhao	4 797	2 533	4 797	100. 0
滨　州 Binzhou	2 180	0	2 099	96. 3

续表 continued

沿海地带 Coastal County (District)	工业废水排放总量 Total Volume of Industrial Waste Water Discharged	工业废水中 Including		
		直接排入海的 Discharged Directly to Sea	符合排放标准的 Up-to-Standard	达标率（%） Standardized Rate（%）
广 东 Guangdong	**90 804**	**3 350**	**82 585**	**90.9**
广 州 Guangzhou	12 266	0	11 747	95.8
深 圳 Shenzhen	6 396	844	6 156	96.2
珠 海 Zhuhai	3 434	328	3 301	96.1
汕 头 Shantou	5 442	346	5 223	96.0
江 门 Jiangmen	8 432	241	7 594	90.1
湛 江 Zhanjiang	6 542	298	4 776	73.0
茂 名 Maoming	5 008	1 013	4 414	88.1
惠 州 Huizhou	2 937	0	2 861	97.4
汕 尾 Shanwei	3 732	2	2 501	67.0
阳 江 Yangjiang	532	26	416	78.2
东 莞 Dongguan	24 134	0	22 344	92.6
中 山 Zhongshan	10 284	250	9 842	95.7
潮 州 Chaozhou	839	0	741	88.3
揭 阳 Jieyang	826	2	668	80.9
广 西 Guangxi	**5 389**	**737**	**4 842**	**89.8**
北 海 Beihai	2 198	25	1 956	89.0
防城港 Fangchenggang	558	432	454	81.4
钦 州 Qinzhou	2 633	280	2 433	92.4
海 南 Hainan	**4 467**	**2 206**	**4 312**	**96.5**
海 口 Haikou	554	0	547	98.7
三 亚 Sanya	292	0	288	98.6

注：表中数据为沿海地带合计数（表8-7、8-10、8-13同）。

Note:The data in the table are the total numbers for the coastal regions(The same for Tables 8-7,8-10,8-13).

8-8 沿海地区工业固体废物排放、处理及综合利用情况（2006年）

Discharge, Treatment and Utilization of Industrial Solid Wastes by Coastal Regions, 2006

单位：吨 Unit:ton

地 区 Region		工业固体废物排放量 Volume of Industrial Solid Wastes Discharged	工业固体废物处置量 Volume of Industrial Solid Wastes Treated	工业固体废物综合利用量 Volume of Industrial Solid Wastes Muti-Utilized
总 计 Total		**1 122 233**	**120 834 495**	**456 405 054**
环渤海经济区 Round-the-Bohai Sea Economic Zone	**合 计 Total**	**670 636**	**92 608 451**	**254 766 717**
	辽 宁 Liaoning	246 918	60 477 591	49 580 058
	河 北 Hebei	419 579	28 498 629	88 504 608
	天 津 Tianjin	0	203 835	12 712 251
	山 东 Shandong	4 139	3 428 396	103 969 800
长江三角洲经济区 Yangzi River Delta Economic Zone	**合 计 Total**	**54 484**	**5 138 630**	**117 741 075**
	江 苏 Jiangsu	270	2 487 376	69 661 688
	上 海 Shanghai	2 414	1 032 184	19 531 133
	浙 江 Zhejiang	51 800	1 619 070	28 548 254
海峡西岸经济区 Economic Zone on West Side of the Straits	**合 计 Total**	**33 707**	**10 699 555**	**31 054 116**
	福 建 Fujian	33 707	10 699 555	31 054 116
珠江三角洲经济区 Zhujiang River Delta Economic Zone	**合 计 Total**	**135 772**	**3 530 794**	**28 781 631**
	广 东 Guangdong	135 772	3 530 794	28 781 631
环北部湾经济区 Round-the-Beibu Gulf Economic Zone	**合 计 Total**	**227 634**	**8 857 065**	**24 061 515**
	广 西 Guangxi	227 434	8 854 987	22 928 344
	海 南 Hainan	200	2 078	1 133 171

8-9 沿海城市工业固体废物排放、处理、综合利用情况（2006年）

Discharge,Treatment and Utilization of Industrial Solid Wastes by Coastal Cities, 2006

单位：吨 Unit:ton

沿海城市 Coastal City	工业固体废物排放量 Volume of Industrial Solid Wastes Discharged	工业固体废物处置量 Volume of Industrial Solid Wastes Treated	工业固体废物综合利用量 Volume of Industrial Solid Wastes Muti-Utilized
天 津 Tianjin	0	203 835	12 712 251
唐 山 Tangshan	256 261	16 043 732	35 766 235
秦皇岛 Qinhuangdao	0	15 510	5 074 311
沧 州 Cangzhou	0	0	580 919
大 连 Dalian	1 074	475 075	2 169 810
丹 东 Dandong	3 415	68 600	1 956 728
锦 州 Jinzhou	872	1 439 570	1 300 856
营 口 Yingkou	0	1 517 618	1 010 495
盘 锦 Panjin	0	109 095	575 595
葫芦岛 Huludao	54 575	661 169	2 233 128
上 海 Shanghai	2 414	1 032 184	19 531 133
南 通 Nantong	0	50 541	2 997 140
连云港 Lianyungang	0	9 380	2 799 783
盐 城 Yancheng	0	12 999	1 644 890

续表 continued

沿海城市 Coastal City	工业固体废物排放量 Volume of Industrial Solid Wastes Discharged	工业固体废物处置量 Volume of Industrial Solid Wastes Treated	工业固体废物综合利用量 Volume of Industrial Solid Wastes Muti-Utilized
杭　州　Hangzhou	1 500	316 595	5 245 446
宁　波　Ningbo	472	520 808	5 636 732
温　州　Wenzhou	17 485	206 065	1 663 434
嘉　兴　Jiaxing	1 114	45 599	2 705 911
绍　兴　Shaoxing	1 640	237 632	2 394 851
舟　山　Zhoushan	6	4 589	409 809
台　州　Taizhou	324	79 020	1 639 738
福　州　Fuzhou	0	29 210	2 005 037
厦　门　Xiamen	211	21 551	785 924
莆　田　Putian	39	85	321 715
泉　州　Quanzhou	3 655	19 969	2 392 877
漳　州　Zhangzhou	1 074	108 950	2 232 817
宁　德　Ningde	1 777	184 775	403 582
青　岛　Qingdao	510	195 269	6 316 073
东　营　Dongying	38	54 116	1 485 036
烟　台　Yantai	0	975 601	12 026 312
潍　坊　Weifang	0	15	4 491 257
威　海　Weihai	0	1 546	1 305 590
日　照　Rizhao	0	149 420	8 308 766
滨　州　Binzhou	0	42 179	4 330 233

续表 continued

沿海城市 Coastal City	工业固体废物排放量 Volume of Industrial Solid Wastes Discharged	工业固体废物处置量 Volume of Industrial Solid Wastes Treated	工业固体废物综合利用量 Volume of Industrial Solid Wastes Muti-Utilized
广　州 Guangzhou	2 598	428 117	6 054 560
深　圳 Shenzhen	500	174 115	877 721
珠　海 Zhuhai	1 700	32 437	570 268
汕　头 Shantou	628	6 303	732 055
江　门 Jiangmen	1 692	51 958	1 329 148
湛　江 Zhanjiang	12 574	132 529	1 396 457
茂　名 Maoming	3 892	39 490	711 810
惠　州 Huizhou	2 299	84 615	105 646
汕　尾 Shanwei	1 096	9 920	89 163
阳　江 Yangjiang	3 752	23 133	316 962
东　莞 Dongguan	10 086	43 680	2 601 747
中　山 Zhongshan	1 348	42 714	314 773
潮　州 Chaozhou	4 279	104 948	161 184
揭　阳 Jieyang	803	1 361	60 382
北　海 Beihai	1 717	627 631	465 864
防城港 Fangchenggang	3 956	0	599 566
钦　州 Qinzhou	9 020	19 839	550 829
海　口 Haikou	0	334	32 003
三　亚 Sanya	0	320	0

8-10 沿海地带工业固体废物排放、处理、综合利用情况（2006年）

Discharge,Treatment and Utilization of Industrial Solid Wastes by Coastal Counties, 2006

单位：吨 Unit:ton

沿海地带 Coastal County （District）	工业固体废物排放量 Volume of Industrial Solid Wastes Discharged	工业固体废物处置量 Volume of Industrial Solid Wastes Treated	工业固体废物综合利用量 Volume of Industrial Solid Wastes Muti-Utilized
天　津　Tianjin	**0**	**165 073**	**4 713 385**
河　北　Hebei	**0**	**15 682**	**10 499 350**
唐　山　Tangshan	0	172	8 472 700
秦皇岛　Qinhuangdao	0	15 510	1 979 135
沧　州　Cangzhou	0	0	47 515
辽　宁　Liaoning	**15 504**	**2 650 889**	**4 973 180**
大　连　Dalian	1 074	443 412	2 099 487
丹　东　Dandong	0	0	508 675
锦　州　Jinzhou	0	28 690	267 657
营　口　Yingkou	0	1 517 618	669 835
盘　锦　Panjin	0	0	44 141
葫芦岛　Huludao	14 430	661 169	1 383 385
上　海　Shanghai	**2 407**	**732 886**	**16 240 452**
江　苏　Jiangsu	**0**	**29 307**	**4 428 418**
南　通　Nantong	0	8 858	648 717
连云港　Lianyungang	0	8 735	2 767 938
盐　城　Yancheng	0	11 714	1 011 763

续表 continued

沿海地带 Coastal County (District)	工业固体废物排放量 Volume of Industrial Solid Wastes Discharged	工业固体废物处置量 Volume of Industrial Solid Wastes Treated	工业固体废物综合利用量 Volume of Industrial Solid Wastes Muti-Utilized
浙　江　Zhejiang	**19 142**	**862 636**	**12 707 265**
杭　州　Hangzhou	0	94 344	1 895 339
宁　波　Ningbo	472	482 378	4 117 056
温　州　Wenzhou	15 740	149 930	1 536 301
嘉　兴　Jiaxing	1 114	22 130	1 765 113
绍　兴　Shaoxing	1 500	60 244	1 554 606
舟　山　Zhoushan	6	4 589	409 809
台　州　Taizhou	310	49 021	1 429 040
福　建　Fujian	**3 587**	**204 504**	**4 898 048**
福　州　Fuzhou	0	17 212	1 478 030
厦　门　Xiamen	211	21 551	785 924
莆　田　Putian	39	85	321 715
泉　州　Quanzhou	1 270	8 362	1 013 631
漳　州　Zhangzhou	756	102 865	929 590
宁　德　Ningde	1 311	54 429	369 159
山　东　Shandong	**548**	**1 130 463**	**21 769 627**
青　岛　Qingdao	510	20 438	1 480 160
东　营　Dongying	38	54 116	1 485 036
烟　台　Yantai	0	905 363	11 794 756
潍　坊　Weifang	0	0	690 695
威　海　Weihai	0	1 546	1 305 590
日　照　Rizhao	0	149 000	2 773 088
滨　州　Binzhou	0	0	2 240 302

续表 continued

沿海地带 Coastal County (District)	工业固体废物排放量 Volume of Industrial Solid Wastes Discharged	工业固体废物处置量 Volume of Industrial Solid Wastes Treated	工业固体废物综合利用量 Volume of Industrial Solid Wastes Muti-Utilized
广 东 Guangdong	**40 365**	**880 136**	**12 850 790**
广 州 Guangzhou	1 622	327 159	4 449 998
深 圳 Shenzhen	500	174 115	877 721
珠 海 Zhuhai	1 700	32 437	570 267
汕 头 Shantou	628	6 327	732 054
江 门 Jiangmen	1 692	47 354	1 128 526
湛 江 Zhanjiang	12 572	131 361	1 390 332
茂 名 Maoming	2 992	34 289	590 267
惠 州 Huizhou	1 499	34 093	71 794
汕 尾 Shanwei	868	4 520	74 883
阳 江 Yangjiang	3 077	209	19 992
东 莞 Dongguan	10 086	43 680	2 601 747
中 山 Zhongshan	1 348	42 714	314 773
潮 州 Chaozhou	1 383	1 402	9 868
揭 阳 Jieyang	400	476	18 570
广 西 Guangxi	**1 401**	**494 763**	**503 115**
北 海 Beihai	445	474 925	146 253
防城港 Fangchenggang	956	0	72 552
钦 州 Qinzhou	0	19 839	284 311
海 南 Hainan	**200**	**1 507**	**665 651**
海 口 Haikou	0	334	27 540
三 亚 Sanya	0	320	0

8-11 沿海地区污染治理项目情况（2006年）

Discharge,Treatment and Utilization of Pollution Treatment Projects by Coastal Regions, 2006

单位：个 Unit:number

地区 Region		当年安排施工项目 Arranged for Construction in the Year		当年竣工项目 Completed in the Year	
		治理废水 Treatment of Waste Water	治理固体废物 Treatment of Solid Wastes	治理废水 Treatment of Waste Water	治理固体废物 Treatment of Solid Wastes
总计 Total		**3 045**	**356**	**2 616**	**331**
环渤海经济区 Round-the-Bohai Sea Economic Zone	**合计 Total**	**956**	**137**	**817**	**132**
	辽宁 Liaoning	133	18	110	15
	河北 Hebei	128	7	107	7
	天津 Tianjin	63	4	57	4
	山东 Shandong	632	108	543	106
长江三角洲经济区 Yangzi River Delta Economic Zone	**合计 Total**	**979**	**44**	**834**	**39**
	江苏 Jiangsu	349	17	310	15
	上海 Shanghai	97	7	79	6
	浙江 Zhejiang	533	20	445	18
海峡西岸经济区 Economic Zone on West Side of the Straits	**合计 Total**	**308**	**88**	**247**	**80**
	福建 Fujian	308	88	247	80
珠江三角洲经济区 Zhujiang River Delta Economic Zone	**合计 Total**	**582**	**63**	**520**	**59**
	广东 Guangdong	582	63	520	59
环北部湾经济区 Round-the-Beibu Gulf Economic Zone	**合计 Total**	**220**	**24**	**198**	**21**
	广西 Guangxi	187	23	169	20
	海南 Hainan	33	1	29	1

8-12 沿海城市污染治理项目情况（2006年）

Discharge,Treatment and Utilization of Pollution Treatment Projects by Coastal Cities, 2006

单位：个 Unit:number

沿海城市 Coastal City	当年安排施工项目 Arranged for Construction in the Year		当年竣工项目 Completed in the Year	
	治理废水 Treatment of Waste Water	治理固体废物 Treatment of Solid Wastes	治理废水 Treament of Waste Water	治理固体废物 Treatment of Solid Wastes
天　津 Tianjin	63	4	57	4
唐　山 Tangshan	17		14	
秦皇岛 Qinhuangdao	3	1	3	1
沧　州 Cangzhou	3	1	3	1
大　连 Dalian	27	3	25	3
丹　东 Dandong	10	1	10	1
锦　州 Jinzhou	8	1	3	1
营　口 Yingkou	3		3	
盘　锦 Panjin	7	4	4	2
葫芦岛 Huludao	5	1	5	1
上　海 Shanghai	97	7	79	6
南　通 Nantong	70	1	68	1
连云港 Lianyungang	7		7	
盐　城 Yancheng	11		10	

续表 continued

沿海城市 Coastal City	当年安排施工项目 Arranged for Construction in the Year		当年竣工项目 Completed in the Year	
	治理废水 Treatment of Waste Water	治理固体废物 Treatment of Solid Wastes	治理废水 Treament of Waste Water	治理固体废物 Treatment of Solid Wastes
杭　州 Hangzhou	101		88	
宁　波 Ningbo	81	4	74	3
温　州 Wenzhou	30		27	
嘉　兴 Jiaxing	52	5	46	5
绍　兴 Shaoxing	40	3	33	3
舟　山 Zhoushan	6		5	
台　州 Taizhou	64	4	56	4
福　州 Fuzhou	50	2	49	2
厦　门 Xiamen	29	2	26	2
莆　田 Putian	2		2	
泉　州 Quanzhou	103	66	64	61
漳　州 Zhangzhou	44		33	
宁　德 Ningde	4	1	3	1
青　岛 Qingdao	29	4	28	4
东　营 Dongying	26	9	22	7
烟　台 Yantai	23	79	22	79
潍　坊 Weifang	72		66	
威　海 Weihai	15	1	15	1
日　照 Rizhao	5		2	
滨　州 Binzhou	12		8	

续表 continued

沿海城市 Coastal City	当年安排施工项目 Arranged for Construction in the Year		当年竣工项目 Completed in the Year	
	治理废水 Treatment of Waste Water	治理固体废物 Treatment of Solid Wastes	治理废水 Treament of Waste Water	治理固体废物 Treatment of Solid Wastes
广　州 Guangzhou	29	1	29	1
深　圳 Shenzhen	34	1	31	1
珠　海 Zhuhai	33	25	33	25
汕　头 Shantou	26		26	
江　门 Jiangmen	22		21	
湛　江 Zhanjiang	42	6	41	6
茂　名 Maoming	25		24	
惠　州 Huizhou	128	3	78	3
汕　尾 Shanwei	9		8	
阳　江 Yangjiang	44		44	
东　莞 Dongguan	19		16	
中　山 Zhongshan	17	11	17	11
潮　州 Chaozhou	14	1	14	1
揭　阳 Jieyang	29	1	29	1
北　海 Beihai				
防城港 Fangchenggang	6		3	
钦　州 Qinzhou	17		16	
海　口 Haikou	7		6	
三　亚 Sanya				

8-13 沿海地带污染治理项目情况（2006年）

Discharge,Treatment and Utilization of Pollution Treatment Projects by Coastal Counties, 2006

单位：个 Unit:number

沿海地带 Coastal County (District)	当年安排施工项目 Arranged for Construction in the Year		当年竣工项目 Completed in the Year	
	治理废水 Treatment of Waste Water	治理固体废物 Treatment of Solid Wastes	治理废水 Treament of Waste Water	治理固体废物 Treatment of Solid Wastes
天 津 Tianjin	**22**	**3**	**20**	**3**
河 北 Hebei	**13**	**2**	**12**	**2**
唐 山 Tangshan	8		7	
秦皇岛 Qinhuangdao	3	1	3	1
沧 州 Cangzhou	2	1	2	1
辽 宁 Liaoning	**38**	**3**	**34**	**3**
大 连 Dalian	24	2	22	2
丹 东 Dandong	3		3	
锦 州 Jinzhou	1			
营 口 Yingkou	3		3	
盘 锦 Panjin	2		1	
葫芦岛 Huludao	5	1	5	1
上 海 Shanghai	**63**	**7**	**52**	**6**
江 苏 Jiangsu	**70**	**1**	**69**	**1**
南 通 Nantong	55	1	55	1
连云港 Lianyungang	5		5	
盐 城 Yancheng	10		9	

续表 continued

沿海地带 Coastal County (District)	当年安排施工项目 Arranged for Construction in the Year		当年竣工项目 Completed in the Year	
	治理废水 Treatment of Waste Water	治理固体废物 Treatment of Solid Wastes	治理废水 Treament of Waste Water	治理固体废物 Treatment of Solid Wastes
浙　江　Zhejiang	**167**	**10**	**153**	**10**
杭　州　Hangzhou	13		13	
宁　波　Ningbo	52	1	49	1
温　州　Wenzhou	16		15	
嘉　兴　Jiaxing	22	4	19	4
绍　兴　Shaoxing	11	1	9	1
舟　山　Zhoushan	6		5	
台　州　Taizhou	47	4	43	4
福　建　Fujian	**159**	**18**	**115**	**14**
福　州　Fuzhou	39	2	38	2
厦　门　Xiamen	29	2	26	2
莆　田　Putian	2		2	
泉　州　Quanzhou	79	14	40	10
漳　州　Zhangzhou	7		7	
宁　德　Ningde	3		2	
山　东　Shandong	**92**	**91**	**79**	**89**
青　岛　Qingdao	14	2	13	2
东　营　Dongying	26	9	22	7
烟　台　Yantai	21	79	20	79
潍　坊　Weifang	6		5	
威　海　Weihai	15	1	15	1
日　照　Rizhao	2			
滨　州　Binzhou	8		4	

续表 continued

沿海地带 Coastal County (District)	当年安排施工项目 Arranged for Construction in the Year		当年竣工项目 Completed in the Year	
	治理废水 Treatment of Waste Water	治理固体废物 Treatment of Solid Wastes	治理废水 Treament of Waste Water	治理固体废物 Treatment of Solid Wastes
广　东 Guangdong	**355**	**45**	**296**	**45**
广　州 Guangzhou	19		19	
深　圳 Shenzhen	34	1	31	1
珠　海 Zhuhai	33	25	33	25
汕　头 Shantou	26		26	
江　门 Jiangmen	10		10	
湛　江 Zhanjiang	42	6	41	6
茂　名 Maoming	22		21	
惠　州 Huizhou	79	1	29	1
汕　尾 Shanwei	7		6	
阳　江 Yangjiang	25		25	
东　莞 Dongguan	19		16	
中　山 Zhongshan	17	11	17	11
潮　州 Chaozhou	8	1	8	1
揭　阳 Jieyang	14		14	
广　西 Guangxi	**9**		**9**	
北　海 Beihai				
防城港 Fangchenggang	1		1	
钦　州 Qinzhou	8		8	
海　南 Hainan	**18**		**16**	
海　口 Haikou	6		6	
三　亚 Sanya				

8-14 沿海区域海洋类型保护区建设情况（2006年）
Construction of Marine-Type Nature Reserves, 2006

地　区 Region	保护区数量（个） Number of Nature Reserves		按保护护级别分（个） By Level of Protection		按保护类型分（个） By Type of Protection			保护区面积(平方公里) Area(km^2)
	已建 Already Established	新建 Newly Established	国家级 Nationed	地方级 Provinced	海洋和海岸生态系统 Marine and Coastal Ecosystem	海洋自然历史遗迹 Marine Natural and Historical Relies	海洋生物多样性 Marine Biodiversity	
合 计 Total	**150**	**2**	**29**	**121**	**90**	**14**	**46**	**34 463.84**
环渤海经济区 Round-the-Bohai Sea Economic Zone	32	1	10	22	20	5	7	19 218.60
长江三角洲经济区 Yangzi River Delta Economic Zone	11	0	5	6	9	0	2	7 367.43
海峡西岸经济区 Economic Zone on West Side of the Straits	15	0	3	12	9	2	4	1 592.41
珠江三角洲经济区 Zhujiang River Delta Economic Zone	60	1	4	56	29	3	28	4 119.03
环北部湾经济区 Round-the-Beibu Gulf Economic Zone	32	0	7	25	23	4	5	2 166.37

8-15 沿海地区海洋类型自然保护区建设情况（2006年）
Construction of Marine-Type Nature Reserves, 2006

地 区 Region	保护区数量（个） Number of Nature Reserves		按保护级别分（个） By Level of Protection		按保护类型分（个） By Type of Protection			保护区面积(平方公里) Area(km^2)
	已建 Already Established	新建 Newly Established	国家级 Nationed	地方级 Provinced	海洋和海岸生态系统 Marine and Coastal Ecosystem	海洋自然历史遗迹 Marine Natural and Historical Relies	海洋生物多样性 Marine Biodiversity	
合 计 Total	**150**	**2**	**29**	**121**	**90**	**14**	**46**	**34 463.84**
天 津 Tianjin	2	0	1	1	1	1	0	1 432.40
河 北 Hebei	4	0	1	3	3	1	0	403.07
辽 宁 Liaoning	14	0	5	9	7	3	4	14 176.70
上 海 Shanghai	4	0	2	2	3	0	1	982.75
江 苏 Jiangsu	4	0	2	2	3	0	1	5 038.97
浙 江 Zhejiang	3	0	1	2	3	0	0	1 345.71
福 建 Fujian	15	0	3	12	9	2	4	1 592.41
山 东 Shandong	12	1	3	9	9	0	3	3 206.43
广 东 Guangdong	60	1	4	56	29	3	28	4 119.03
广 西 Guangxi	6	0	3	3	5	0	1	806.78
海 南 Hainan	26	0	4	22	18	4	4	1 359.59

注：全国海洋特别保护区8个，江苏、广东各1个，浙江、福建各3个。

Note:There are eight national special marine protected areas, i.e,1 in Jiangsu,1 in Guangdong,3 in Zhejiang and 3 in Fujian respectirely.

8-16 全国海洋生态监控区基本情况（2006年）
Basic Condition of the Marine Ecological Monitoring Areas Throughout the Country, 2006

生态监控区 Ecological Monitoring Area	所在地 Location	面积 (平方公里) Area (km^2)	主要生态系统类型 Major Types of Ecosystem	健康状况 Health Condition	三年变化趋势 Three-year variation trend
双台子河口	辽宁省	3 000	河 口	亚健康	略有好转
Shuangtaizi Estuary	Liaoning Province		Estuary	Subhealth	Slightly Improved
锦州湾*	辽宁省	650	海 湾	不健康	基本稳定
Jinzhou Bay	Liaoning Province		Bay	Poor Health	Basically Stable
滦河口-北戴河	河北省	900	河 口	亚健康	基本稳定
Luanhe Mouth-Beidaihe	Hebei Province		Estuary	Subhealth	Basically Stable
渤海湾	天津市	3 000	海 湾	亚健康	基本稳定
Bohai Bay	Tianjin Municipality		Bay	Subhealth	Basically Stable
莱州湾	山东省	3 770	海 湾	不健康	略有好转
Laizhou Bay	Shandong Province		Bay	Poor Health	Slightly Improved
黄河口	山东省	2 600	河 口	亚健康	略有好转
Yellow River Mouth	Shandong Province		Estuary	Subhealth	Slightly Improved
苏北浅滩	江苏省	3 090	湿 地	亚健康	基本稳定
North Jiangsu Bank	Jiangsu Province		Wetland	Subhealth	Basically Stable
长江口	上海市	13 668	河 口	亚健康	略有好转
Changjiang River Mouth	Shanghai Municipality		Estuary	Subhealth	Slightly Improved
杭州湾	上海市 浙江省	5 000	海 湾	不健康	基本稳定
Hangzhou Bay	Shanghai Municipality Zhejiang Province		Bay	Poor Health	Basically Stable
乐清湾	浙江省	464	海 湾	亚健康	基本稳定
Leqing Bay	Zhejiang Province		Bay	Subhealth	Basically Stable

续表 continued

生态监控区 Ecological Monitoring Area	所在地 Location	面积 (平方公里) Area (km^2)	主要生态系统类型 Major Types of Ecosystem	健康状况 Health Condition	三年变化趋势 Three-year variation trend
闽东沿岸	福建省	5 063	海 湾	亚健康	基本稳定
Constal East Fujian	Fujian Province		Bay	Subhealth	Basically Stable
大亚湾	广东省	1 200	海 湾	亚健康	基本稳定
Daya Bay	Guangdong Province		Bay	Subhealth	Basically Stable
珠江口	广东省	3 980	河 口	不健康	基本稳定
Zhujiang River Mouth	Guangdong Province		Estuary	Poor Health	Basically Stable
雷州半岛 西南沿岸	广东省	1 150	珊瑚礁	亚健康	略有下降
Southwest Coast of Leizhou Peninsula	Guangdong Province		Coral Reef	Subhealth	Slightly Declining
广西北海	广西壮族自治区	120	珊瑚礁、红树林、海草床	健 康	基本稳定
Beihai,Guangxi	Guangxi Zhuang Nationality Autonomous Region		Coral Reef,Mangroves,Seagrass Bed	Healthy	Basically Stable
北仑河口*	广西壮族自治区	150	红树林	健 康	基本稳定
Beilun River Mouth	Guangxi Zhuang Nationality Autonomous Region		Mangroves	Healthy	Basically Stable
海南东海岸	海南省	3 750	珊瑚礁、海草床	健 康	基本稳定
East Coast of Hainan	Hainan Province		Coral Reef,Seagrass Bed	Healthy	Basically Stable
西沙珊瑚礁*	海南省	400	珊瑚礁	健 康	基本稳定
Xisha Coral Reef	Hainan Province		Coral Reef	Healthy	Basically Stable

注：2005年新增生态监控区，变化趋势指两年。
Note:The newly added ecological monitoring areas with two-year variation trends.

8-17 沿海地区风暴潮灾害情况(2006年)
Survey of Storm Surges by Coastal Regions, 2006

受灾地区 Disastered Region	受灾人口(万人) Affected Population (10 000 persons)	农作物受灾(万公顷) Inundated Area of Farmland (10 000 hm²)	损失船只(艘) Lost Boat (Number)	海水养殖受灾面积(万公顷) Affected Area of Mariculture (10 000 hm²)
合 计 Total	**2 688.30**	**95.4**	**6 980**	**12.14**
天 津 Tianjin				
河 北 Hebei				
辽 宁 Liaoning				5 160(吨)
上 海 Shanghai				
江 苏 Jiangsu				
浙 江 Zhejiang	520.40	15.80	2 510	1.52
福 建 Fujian	748.40	24.50	2 444	0.60
山 东 Shandong				
广 东 Guangdong	1 251.70	55.10	2 026	9.45
广 西 Guangxi	167.80			0.57
海 南 Hainan				

续表 continued

受灾地区 Disastered Region	毁损、决口海塘堤防(处、公里) Destroyed Seawall and Dyke breaching(km)	损毁海洋工程(座) Destroyed Marine Engineering Architectures	死亡人数 (人) Number of Deaths (person)	经济损失(亿元) Economic Loss (100 million yuan)
合 计 Total	**堤防毁损7949处，711公里 Destroyed Seawall: 7949 sites, 711 km**		**327**	**217.11**
天 津 Tianjin				
河 北 Hebei				
辽 宁 Liaoning				0. 03
上 海 Shanghai				
江 苏 Jiangsu				
浙 江 Zhejiang	堤防毁损6274处，530. 4公里 Destroyed Seawall 6274 sites, 530.4 km		2	13. 23
福 建 Fujian	不详		324	114. 49
山 东 Shandong				
广 东 Guangdong	堤防毁损1675处，144.89公里 Destroyed Seawall 1675 sites, 144.89 km			82. 32
广 西 Guangxi	堤防毁损35.7公里 Destroyed Seawall 35.7 km		1	7. 04
海 南 Hainan				

8-18 沿海地区赤潮灾害情况（2006年）
Survey of Red Tide by Coastal Regions, 2006

时 间 Date	地 区 Area		受灾面积(平方公里) Disaster Area(km^2)
总 计 Total			**10 555**
其中: Including:			
5月3日-5月6日 May 3-May 6	浙江省舟山外海域	Sea Area Off Zhoushan, Zhejiang Province	1 000
5月14日 May 14	长江口外海域	Sea Area off the Changjiang River Mouth	1 000
5月20日-5月27日 May 20–May 27	浙江渔山列岛附近海域	Sea Area Around the Yushan Arclipelago, Zhejiang	3 000
6月12日-6月14日 Jun 12–Jun 14	浙江省洞头岛到南麂列岛附近海域	From Dongtou Island to the Sea Area Around the Nanji Archipelago,Zhejiang Province	2 100
6月24日-6月27日 Jun 24–Jun 27	浙江省渔山列岛到韭山列岛附近海域	From Yushan Archipelago to the Sea Area Around the Jiushan Archipelago	1 200
10月2日-10月7日 Oct 2–Oct 7	江苏省连云港市海州湾海域	Sea Area of Haizhou Bay,Lianyungang City Jiangsu Province	400
10月8日-10月26日 Oct 8–Oct 26	天津近岸海域	Nearshore Sea Area of Tianjin	210
10月22日-11月5日 Oct 22–Nov 5	河北省黄骅附近	Around Huanghua,Hebei Province	1 600
12月3日-12月23日 Dec 3–Dec 23	广东省汕尾港海域	Sea Area of Shanwei Port,Guangdong Province	45

主要统计指标解释

1.工业废水排放量 指经过企业厂区所有排放口排到企业外部的工业废水量。包括生产废水、外排的直接冷却水、超标排放的矿井地下水和与工业废水混排的厂区生活污水,不包括外排的间接冷却水(清污不分流的间接冷却水应计算在内)。

2.直接排入海的工业废水量 指经企业位于海边的排放口,直接排入海中的废水量。直接排入是指废水经过工厂的排污口直接排入海,而未经过城市下水道或其他中间体,也不受其他水体的影响。

3.工业废水处理量 指报告期内各种水治理设施实际处理的工业废水量,包括处理后外排的和处理后回用的工业废水量,虽经处理但未达到国家或地方排放标准的废水量也应计算在内。计算时,如遇车间和厂排放口均有治理设施,并对同一废水分级处理时,不应重复计算工业废水处理量。

4.工业废水处理率 指工业废水处理量占需要处理的工业废水量的百分率。其计算公式是:

工业废水处理率 =(工业废水处理量/ 需处理的工业废水量)×100%

式中,需处理工业废水量 =工业废水排放量+工业废水处理回用量-(工业废水排放达标量-工业废水处理排放达标量)

5.工业废水排放达标量 指各项指标都达到国家或地方排放标准的外排工业废水量,包括经过处理后外排达标的和未经处理外排达标的两部分。国家排放标准见 GB 8978-88。

6. 工业废水处理排放达标量 指经过各种水治理设施处理后达到国家或地方排放标准的废水排放量。

7. 工业固体废物处置量 指将固体废物焚烧或者最终置于符合环境保护规定要求的场所并不再回取的工业固体废物量(包括当年处置往年的工业固体废物累计贮存量)。

8.工业固体废物排放量 指将所产生的固体废物排到固体废物污染防治设施、场所以外的量。不包括矿山开采的剥离废石和掘进废石(煤矸石和呈酸性或碱性的废石除外)。

9.当年开工污染治理项目数 指报告期内由国家、部门、地方或企业单位安排开工的,并以治理废水、废气、固体废物、噪声和其他(如电磁波、恶臭等)环境污染的环境治理工程的总数。不包括"三同时"项目。

10.当年竣工项目数 指报告期内竣工投入运行的治理废水、废气、固体废物、噪声及其他污染的环境工程项目的总数。

Explanatory Notes on Main Statistical Indicators

1.Volume of Industrial Waste Water Discharged refers to the quantity of industrial waste water discharged externally through all the outlets in the factory area of the enterprise, including the waste water from production, externally discharged direct cooling water, mine-shaft groundwater discharged exceeding the set standard and the domestic sewage of the factory area discharged together with the industrial waste water, but not including the indirect cooling water discharge externally.

2.Volume of Industrial Waste Water Discharged Directly to Sea refers to the quantity of waste water directly discharged into the sea through the outlets of the enterprise by the sea. Direct discharge means the direct discharge into the sea of waste water through the outlets of the factory, which is not discharged via the urban sewers or other intermediates and is not affected by other water bodies.

3.Volume of Industrial Waste Water Treated refers to the industrial waste water volume actually treated by various water treatment facilities in the period covered by the report, including the quantity of the industrial waste water discharged and reused after treatment. The amount of the waste water which is not up to the state or local standard of discharge upon treatment should be included. If the workshops and outlets of the factory are provided with treatment facilities and carry out graded treatment of the same waste water, the processed volume of industrial waste water cannot be calculated repeatedly.

4.Rate of Industrial Waste Water Treatment refers to the percentage of the processed volume of industrial waste water in the industrial waste water volume that needs to be treated. The calculation formula is:

Rate of Industrial Waste Water Treatment = (Processed Volume of Industrial Waste Water /Processed Volume of Industrial Waste Water Required) × 100%.

Where: Processed Volume of Industrial Waste Water Required = (Processed Volume of Industrial Waste Water + Reused Volume of Industrial Waste Water After Treatment - Up-to-Standard Volume of Industrial Waste Water for Discharge - Up-to-Standard Volume of Industrial Waste Water for Discharge after Treatment).

5. Volume of Up–to–Standard Industrial Waste Water Discharged refers to the externally discharged volume of industrial waste water with all its indexes up to the state or local standard for discharge, including the volume of industrial waste water up to the standard for discharge after being treated or that without being treated. See GB8978-88 for the state's discharge standard.

6.Volume of Up–to–Standard Discharged Industrial Waste Water After Treatment refers to the discharged volume of waste water that is up to the state or local standard for discharge upon treatment by various water treatment facilities.

7.Volume of Industrial Solid Wastes Discharged refers to the amount of solid wastes discharged out of the facilities and sites for the solid wastes pollution prevention and control, not including the stripped and tunneled waste ores in the excavation of mines (other than gangues and acid or alkaline waste ores).

8.Volume of Industrial Solid Wastes Treated refers to the volume of industrial solid wastes which are to be burned or finally placed at the sites in keeping with the requirement of environmental protection and will not be recovered (including the accumulated amount of storage in former years of industrial solid radioactive matter disposed of in the year).

9.Number of Pollution Treatment Projects Started in the Current Year refers to the total number of environmental pollution control projects for controlling waste water, waste gas, solid wastes, noise and other environmental pollutions (such as electromagnetic wave, offensive odor) started by the state, governmental departments, local governments or enterprises in the period covered by the report mainly for the purpose of pollution control and multipurpose utilization of "three wastes".

10.Number of Pollution Treatment Projects Completed in the Current Year refers to the total number of environmental engineering projects for controlling waste water, waste gas, solid wastes, noise and other pollutions which are completed and put into operation in the period covered by the report.

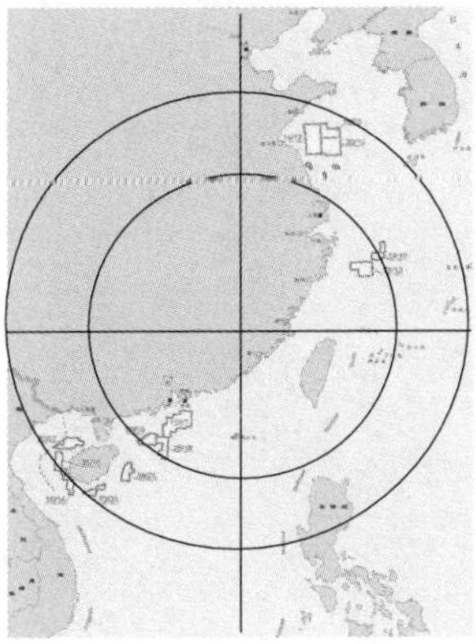

9

海洋行政管理及公益服务

Marine Administration and Public–Good Service

9-1 海域使用管理情况（2006年）
Sea Area Use Management, 2006

部门和地区 Sector and Region	发放海域使用权证书（本） Certificates of Right of Sea Area Use Issued	确权海域面积（公顷） Area of Waters with Established Rights (hm^2)	海域使用金(万元) Charge for Sea Area Utilization (10 000 yuan)
合 计 **Total**	**8 759**	**227 318**	**157 474.53**
国家级 National-Level	38	8 588	50 341.51
天 津 Tianjin	26	1 120	1 389.52
河 北 Hebei	1 412	21 904	11 314.56
辽 宁 Liaoning	2 392	54 682	22 629.15
上 海 Shanghai	5	241	71.95
江 苏 Jiangsu	378	61 788	14 423.08
浙 江 Zhejiang	323	7 969	8 771.25
福 建 Fujian	1 312	11 004	13 637.08
山 东 Shandong	1 806	47 836	22 974.72
广 东 Guangdong	453	9 585	2 132.40
广 西 Guangxi	297	1 328	8 721.02
海 南 Hainan	317	1 274	1 068.29

9-2 海洋倾废管理情况（2006年）
Management on the Ocean Dumping of Wastes, 2006

地 区 Region	签发疏浚物海洋倾倒许可证（份） Permit Issued for Ocean Dumping of Dredged Materials (number)	倾倒区情况 Condition of Dumping Zones	
		现有数量（个） Available Number (number)	现有面积（平方公里） Available Area (km^2)
合 计 Total	**466**	**83**	**352.62**
天 津 Tianjin	21	3	7.07
河 北 Hebei	28	6	24.34
辽 宁 Liaoning	26	10	15.70
上 海 Shanghai	190	11	24.99
江 苏 Jiangsu	15	2	4.45
浙 江 Zhejiang	50	11	16.94
福 建 Fujian	27	7	36.97
山 东 Shandong	63	12	36.30
广 东 Guangdong	38	16	152.84
广 西 Guangxi	2	1	25.40
海 南 Hainan	6	4	7.62

9-3 海洋执法检查情况（2006年）
Marine Law Enforcement Inspection, 2006

执法检查项目分类 Classification of Items Subject to Low Enforcement Inspection	检查项目(个) Number of Items Subject to Inspection	检查次数(次) Number of Times of Inspection	发现违法行为(起) Illegal Acts Found (Cases)
合　计 Total	**19 710**	**45 088**	**3 101**
海域使用 Sea Area Use	18 229	36 232	2 436
海底电缆管道 Submarine Cable Pipeline	99	1 185	17
海洋工程建设项目 Marine Engineering Construction Projects	357	1 649	52
海洋倾废 Ocean Waste-Dumping	607	2 146	192
海洋生态保护 Marine Ecological Protection	418	3 876	404

9-4 沿海地区海滨观测台站分布概况（2006年）

Distribution of Coastal Observation Stations by Coastal Regions,2006

单位：个 Unit: number

地 区 Region	合 计 Total	海洋站 Marine Station	验潮站① Tide Station	气象台站② Meteorological Station	地震台站 Seismic Station	雷达站 Radar Station
合 计 Total	**789**	**66**	**232**	**100**	**334**	**57**
天 津 Tianjin	**44**	1	5	2	34	2
河 北 Hebei	**38**	1		4	28	5
辽 宁 Liaoning	**52**	7	2	7	30	6
上 海 Shanghai	**112**	5	71	1	23	12
江 苏 Jiangsu	**58**	4	19	19	14	2
浙 江 Zhejiang	**88**	9	37	20	16	6
福 建 Fujian	**115**	7	11	27	67	3
山 东 Shandong	**62**	8	3	8	35	8
广 东 Guangdong	**165**	13	77	7	60	8
广 西 Guangxi	**12**	3	4	3	2	0
海 南 Hainan	**42**	8	3	1	25	5

注： ①潮流量观测站39处，潮水位观测站193处。
②中国气象局所属的气象台站仅指沿海气象台站中有海洋气象观测、预报及服务等业务的台站，总计中含有中央气象台1个。

Note: ①There are 39 tidal current observation stations and 193 tidal level observation stations.
② The meteorological observatories and stations under the China Meteorological Administration only refer to those with marine meteorotogical observation, forecast and service among the coastal meteorological observatories and the total includes the central meteorological station.

9-5 海洋预报服务概况（2006年）

Marine Forecast Service, 2006

预报项目 Item	数值预报 Numerical Forecast			
	预报服务次数 Frequency	发布次数 Frequency of Release		
		广播电视 Radio and TV	因特网 Internet	纸 质 Paper Media
合 计 Total	**943 085**	**16 865**	**946 849**	**32 831**
海 浪 Wave	18 216	5 430	22 005	8 001
海 温 SeaLevel Temperature	188 442	4 700	194 307	6 449
潮 汐 Tide	22 314	5 765	17 050	6 509
海 流 Sea Current	175 646		175 727	162
海平面 Sea Level	72		72	
盐 度 Salinity	175 214		175 200	
赤 潮 Red Tide				
海水浴场 Bathing Beach	645	889	732	488
海 冰 Sea Ice				
风暴潮 Storm Surge	1 005	81	3 623	1 033
气 象 Metorology	7 988		7 022	7 136
厄尔尼诺 EL Nino	9		27	377
专 项 Special Item	3 110		664	2 652
其 他 Others	350 424		350 420	24

续表 continued

预报项目 Item	统计预报 Statistical Forecast			
	预报服务次数 Frequency	发布次数 Frequency of Release		
		广播电视 Radio and TV	因特网 Internet	纸 质 Paper Media
合 计 Total	**368 212 540**	**26 794**	**366 682 211**	**1 550 395**
海 浪 Wave	22 621	9 031	25 483	10 342
海 温 SeaLevel Temperature	17 655	7 996	23 422	8 720
潮 汐 Tide	47 238 298	7 354	45 836 677	1 407 788
海 流 Sea Current	320 860 266		320 765 447	95 062
海平面 Sea Level	105		105	
盐 度 Salinity				
赤 潮 Red Tide	939	368	1 613	1 242
海水浴场 Bathing Beach	2 356	1 522	1 894	1 301
海 冰 Sea Ice	279	20	35	104
风暴潮 Storm Surge	997	314	3 019	3 649
气 象 Metorology	24 813		8 325	8 813
厄尔尼诺 EL Niño	9		27	377
专 项 Special Item	43 391		15 820	12 807
其 他 Others	811	189	344	190

注:本表只包含国家海洋局资料。

Note:This table only contains the data from the State Oceanic Administration.

9-6 海洋监测情况（2006年）
Marine Monitoring Statistics, 2006

项目 Item	合　计 Total	台站监测 Station Monitoring	污染监测 Pollution Monitoring	浮标监测 Buoy Monitoring	断面监测 Sectional Monitoring	其他监测 Other Monitoring
测站（点）数（个） Number of Station	**9 469**	52	8 800	3	170	444
实际获得数据量(万个) Quantity of Data Actually Obtained (10 000 units)	**7 522.81**	6 385.80	180.00	927.17	1.89	27.95

9-7 海洋调查概况（2006年）
Marine Survey Statistics, 2006

调查名称 Name	站点数（个） Number of Stations	船舶数（艘） Number of Ships	项目数（个） Number of Items	实际获得数据（个） Quantity of Data Actually Obtained	发布通（公、简）报量（期） Quantity of Circulars (Bulletin, Brief Reports)	提交报告数（份） Number of Reports Submitted
合　计 Total	**5 988**	**188**	**836**	**267 865**	**275**	**127**
大洋调查 Oceanic Survey	704	2	18	796	199	34
极地调查 Polar Survey	69	1	18	15 009	1	5
专项调查 Special Survey	3 787	63	256	47 039	39	66
其他调查 Other Survey	1 428	122	544	205 021	36	22

9-8 海洋档案及利用情况（2006年）
Marine File and Its Use, 2006

指　标 Item	指标值 Data
室（馆）存档案 Files Deposited in the Archives	
纸介质（卷、册） Paper Media (reel,volume)	105 918
磁介质（盘） Magnetic Media (spool)	6 215
科技档案利用 Utilization of Scientific as Technological Files	
接待读者（人次） Readers Received (person-time)	2 035
借阅案卷、盘 Borrowed Files and Spools	
（卷次） Number of Reels	3 749
（盘次） Number of Spools	81

9-9 卫星遥感接收应用情况（2006年）
Remote-Sensing Receiving and Utilization ,2006

指 标 Item	指标值 Data
卫星接收次数（次） Number of Satellite Receptions (time)	112 984
全年接收时间（分钟） Whole Year Receiving Time (minute)	6 432 755
全年实际接收存档数据量（GB） Amount of Data Actually Received and Placed on File in the whole year (GB)	12 345.79
累计存档数据量（GB） Total Amount of Data Placed on File (GB)	26 601.88
卫星数据分发 Satellite Data Distribution	
类别用户（个） Classified Users(number)	87
分发数据量（GB） Amount of Data Distributed(GB)	5 079.04

9-10 海洋标准化监督管理情况（2006年）
Supervision and Management of Marine Standardizalion, 2006

指　标 Item	指标值 Data
标准立项审查 Examination of the Standards for Authorization	
国家标准（项） National Standards	18
行业标准（项） Professional Standards	28
标准监督审查 Supervision and Examination of Standards	
国家标准（项） National Standards	27
行业标准（项） Professional Standards	6
标准出版 Standards Publication	
国家标准（项） National Standards	6
行业标准（项） Professional Standards	1
标准实施监督检查（次） Implementation Supervision and Examination of Standards (time)	10

主要统计指标解释

1.海底电缆管道检查 主要是针对铺设海底电缆管道路由调查、铺设施工和维修改造等的监督检查。

2. 海洋工程建设项目环境保护检查 主要是针对防治海洋工程建设项目对海洋环境的污染损害的监督检查。

3.海洋倾废检查 主要是针对防治倾倒废弃物对海洋环境的污染损害的监督检查。

4.海洋生态保护检查 主要是针对红树林、珊瑚礁、滨海湿地、海岛、海湾、入海河口、重要渔业水域等具有典型性、代表性的海洋生态系统，珍稀、濒危海洋生物的天然集中分布区，具有重要经济价值的海洋生物生存区域及有重大科学文化价值的海洋自然历史遗迹和自然景观等海洋自然保护区以及其他需要予以特殊保护的区域的监督检查。

5.断面监测 按照国家海洋局"断面监测方案"，每年定期利用船舶在沿海设定的断面上进行海洋水文、气象、生物、化学等项目的监测活动。

6.浮标监测 在海上固定站位获取长期、连续海洋环境观测资料的海上锚定资料浮标。

7.大洋调查 以大洋科考、研究为目的的远洋调查。

8.专项调查 为完成国家专项任务进行的海洋调查。

9.全年接收时间 指全年整个应用系统接收的各类遥感卫星数据时，接收设备工作时间。

10.全年实际接收存档数据量 指全年整个应用系统接收的各类遥感卫星数据原始数据量。

11.累计存档数据量 指全年整个应用系统制作的各类遥感产品数据量。

12.分发数据量 指应用系统提供数据产品用于开展各类应用的数据量。

13.卫星接收次数 地面观测系统接收卫星数据的数量。

14. 国家标准 针对海洋领域内需要在全国范围内统一的有关技术要求所制定的国家标准。海洋国家标准由国家标准化主管部门统一批准、编号和发布。

15. 行业标准 对没有海洋国家标准而又需要在海洋领域内统一的技术要求所制定的标准。海洋行业标准由国家海洋局统一批准、编号和发布。

16.获奖成果数 指本年度内机构作为第一完成单位从地(市)以上政府科技管理部门获得的各种科技成果奖的项数。

17.国家级奖励 指国家自然科学奖、国家发明奖、国家科技进步奖。

18.省部级奖励 指以国务院各部门名义颁发的或以省、自治区、直辖市政府(科委)名义颁发的重大科技成果奖和科技进步奖等。

19.地市级奖励 指以地(市、州)政府(科委)和省属各部门名义颁发的重大科技成果奖和科技进步奖等。

Explanatory Notes on Main Statistical Indicators

1.Inspection of Submarine Cables and Pipelines is mainly aimed at the survey of the submarine cables and pipelines laying route and supervision and inspection of the laying construction, repair and transformation.

2.Environmental Protection Inspection for the Marine Engineering Construction Projects is mainly directed against preventing and controlling the pollution damage of the marine engineering construction project to the marine environment.

3.Inspection of Oceanic Dumping of Wastes mainly refers to the supervision and inspection aimed at preventing and controlling the pollution damage of wastes dumping to the marine environment.

4.Inspection of Marine Ecological Protection mainly refers to the supervision and inspection of the typical and representative marine ecosystems such as mangroves, coral reef, coastal wetland, sea island, bay, estuaries open to the sea, important fishery waters, etc, the natural centralized distribution zones of rare and endangered marine life, the living areas for the marine life with significant economic values and the marine nature reserves such as the marine natural and historical remains and natural landscapes with important scientific and cultural values as well as other areas needing to be specially protected.

5.Sectional Monitoring According to the "Sectional Monitoring Plan" of the State Oceanic Administration, monitoring activities concerning such items as marine hydrology, meteorology, biology and chemistry are carried out regularly every year on the sections set in the coastal area.

6.Buoy Monitoring refers to the monitoring carried out by the offshore mooring data buoys which acquire long-term, continuous marine environmental observations at the fixed stations at sea.

7.Oceanic Survey refers to the oceanic surveys aimed at the oceanic scientific investigations and research.

8.Special–Subject Survey refers to the oceanic investigation for the purpose of fulfilling the state's special tasks.

9. Whole Year Receiving Time refers to the operating time of receiving equipment in the whole year when the whole application system receives all types of remote sensing satellite data.

10.Amount of Data Actually Received and Placed on File Throughout the Year refers to the raw data of remote-sensing satellite data of various kinds received by the whole application system throughout the year.

11.Total Amount of Data Placed on File refers to the data amount of various remote-sensing products made by the whole application systems throughout the year.

12.Amount of Data Distributed refers to the amount of data products provided by the application system for various uses.
13.Number of Satellite Receptions refers to the amount of data received by the ground observation system.
14.National Standards refers to the standards formulated in view of the relevant technical requirements in the marine field that need to be unified throughout the country. The marine national standards are approved, numbered and issued uniformly by the state department responsible for standardization.
15 Professional Standards refers to the standards formulated for the technical requirements which have no national standards but need to be unified in the marine field. The Marine Professional Standards are approved, numbered and issued by the State Oceanic Administration.
16.Number of Prize–Winning Results refers to the number of items of various scientific and technological results awards obtained from the departments of scientific and technological management of the governments above the prefecture (city) level by the institution and the first accomplishing unit.
17.State–Level Awards refers to the State Natural Science Award, State Invention Prize and State Scientific and Technological Progress Prize.
18.Province (Ministry)–Level Awards refers to the significant scientific and technological achievement prize and the scientific and technological progress prize awarded in the name of the departments of the State Council or the governments of provinces, autonomous regions and municipalities directly under the Central Government (commissions of science and technology).
19.Prefecture–Level Awards refers to the significant scientific and technological achievements prize and the scientific and technological progress prize awarded in the name of the prefecture government (commission of science and technology) and the various departments of the province.

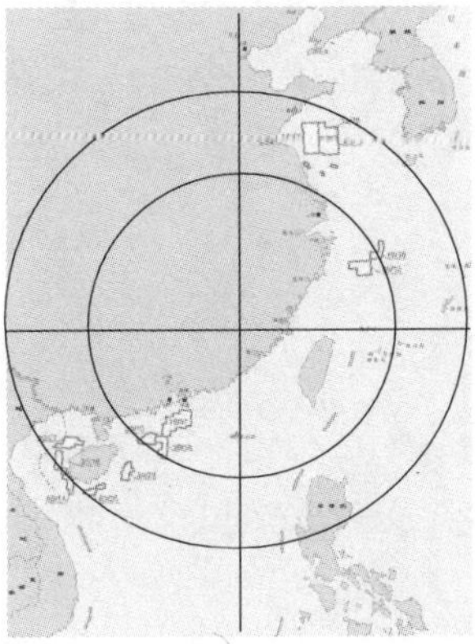

10

全国及沿海社会经济

National and Coastal Socioeconomy

10-1 国内生产总值
Gross Domestic Product

单位：亿元 Unit:100 million yuan

年 份 Year	国内生产总值 Gross Domestic Product			
		第一产业 Primary Industry	第二产业 Secondary Industry	第三产业 Tertiary Industry
2001	109 655	15 781	49 512	44 362
2002	120 333	16 537	53 897	49 899
2003	135 823	17 382	62 436	56 005
2004	159 878	21 413	73 904	64 561
2005	183 868	23 070	87 365	73 433
2006	210 871	24 737	103 162	82 972

10-2 国内生产总值增长速度
Growth Rate of Gross Domestic Product

年 份 Year	国内生产总值(%) Gross Domestic Product(%)			
		第一产业 Primary Industry	第二产业 Secondary Industry	第三产业 Tertiary Industry
2001	8. 3	2. 8	8. 4	10. 3
2002	9. 1	2. 9	9. 8	10. 4
2003	10. 0	2. 5	12. 7	9. 5
2004	10. 1	6. 3	11. 1	10. 1
2005	10. 4	5. 2	11. 7	10. 5
2006	11. 1	5. 0	13. 0	10. 8

注：本表按可比价格计算(上年为基期)。

Note:This table is calculated at the comparable price (with the previons year as the base period).

10-3 沿海地区生产总值(2006年)

Gross Regional Product of Coastal Regions, 2006

单位：亿元 Unit:100 million yuan

地　区 Region	地区生产总值 Gross Regional Product	第一产业 Primary Industry	第二产业 Secondary Industry	第三产业 Tertiary Industry
合　计 Total	**134 802.43**	**11 254.13**	**71 214.73**	**52 333.57**
天　津 Tianjin	4 359.15	118.23	2 488.29	1 752.63
河　北 Hebei	11 660.43	1 606.48	6 115.01	3 938.94
辽　宁 Liaoning	9 251.15	976.37	4 729.50	3 545.28
上　海 Shanghai	10 366.37	93.80	5 028.37	5 244.20
江　苏 Jiangsu	21 645.08	1 545.01	12 250.84	7 849.23
浙　江 Zhejiang	15 742.51	925.10	8 509.57	6 307.85
福　建 Fujian	7 614.55	896.17	3 743.71	2 974.67
山　东 Shandong	22 077.36	2 138.90	12 751.20	7 187.26
广　东 Guangdong	26 204.47	1 577.12	13 431.82	11 195.53
广　西 Guangxi	4 828.51	1 032.47	1 878.56	1 917.47
海　南 Hainan	1 052.85	344.48	287.86	420.51

10-4 沿海地区生产总值增长速度(%)

Growth Rate of Gross Regional Product of Coastal Regions(%)

地 区 Region	2001	2002	2003	2004	2005	2006
天 津 Tianjin	12.0	12.7	14.8	15.8	14.7	14.5
河 北 Hebei	8.7	9.6	11.6	12.9	13.4	13.4
辽 宁 Liaoning	9.0	10.2	11.5	12.8	12.3	13.8
上 海 Shanghai	10.5	11.3	12.3	14.2	11.1	12.0
江 苏 Jiangsu	10.2	11.7	13.6	14.8	14.5	14.9
浙 江 Zhejiang	10.6	12.6	14.7	14.5	12.8	13.9
福 建 Fujian	8.7	10.2	11.5	11.8	11.6	14.8
山 东 Shandong	10.0	11.7	13.4	15.4	15.2	14.8
广 东 Guangdong	10.5	12.4	14.8	14.8	13.8	14.6
广 西 Guangxi	8.3	10.6	10.2	11.8	13.2	13.6
海 南 Hainan	9.1	9.6	10.6	10.7	10.2	12.5

注：本表按可比价格计算(上年为基期)。

Note:This table is calculated at the comparable price (with the previons year as the base period).

10-5 沿海城市生产总值(2005年)

Gross Regional Product of Coastal Cities, 2005

单位：亿元 Unit:100 million yuan

沿海城市 Coastal City		地区生产总值 Gross Regional Product	第一产业 Primary Industry	第二产业 Secondary Industry	第三产业 Tertiary Industry
合　计	**Total**	**69 203.90**	**4 871.00**	**35 797.70**	**28 256.30**
天　津	**Tianjin**	**3 697.60**	**112.40**	**2 051.20**	**1 534.10**
河　北	**Hebei**	**3 649.60**	**422.90**	**1 955.90**	**1 270.80**
唐　山	Tangshan	2 027.60	236.20	1 161.70	629.70
秦皇岛	Qinhuangdao	491.20	51.30	190.40	249.50
沧　州	Cangzhou	1 130.80	135.40	603.80	391.60
辽　宁	**Liaoning**	**3 980.80**	**456.60**	**1 926.60**	**1 597.70**
大　连	Dalian	2 152.20	183.30	996.10	972.80
丹　东	Dandong	326.30	52.80	134.20	139.30
锦　州	Jinzhou	381.90	88.00	140.10	153.80
营　口	Yingkou	379.60	41.00	195.00	143.60
盘　锦	Panjin	441.30	46.90	318.80	75.70
葫芦岛	Huludao	299.50	44.60	142.40	112.50
上　海	**Shanghai**	**9 154.20**	**80.30**	**4 452.90**	**4 620.90**
江　苏	**Jiangsu**	**2 933.00**	**481.90**	**1 469.60**	**981.50**
南　通	Nantong	1 472.10	160.00	825.90	486.20
连云港	Lianyungang	456.00	93.40	201.50	161.10
盐　城	Yancheng	1 004.90	228.50	442.20	334.20
浙　江	**Zhejiang**	**11 127.60**	**665.40**	**6 029.60**	**4 432.50**
杭　州	Hangzhou	2 942.70	148.20	1 496.90	1 297.50
宁　波	Ningbo	2 449.30	132.30	1 341.50	975.60
温　州	Wenzhou	1 596.40	64.90	866.90	664.60
嘉　兴	Jiaxing	1 159.70	84.50	681.90	393.30
绍　兴	Shaoxing	1 447.50	93.10	873.00	481.40
舟　山	Zhoushan	280.20	39.80	111.30	129.10
台　州	Taizhou	1 251.80	102.60	658.10	491.00
福　建	**Fujian**	**5 441.20**	**583.80**	**2 751.90**	**2 105.50**
福　州	Fuzhou	1 476.30	174.80	693.90	607.60
厦　门	Xiamen	1 006.60	21.00	552.30	433.30
莆　田	Putian	359.90	51.40	192.00	116.50
泉　州	Quanzhou	1 626.30	97.90	939.50	588.90
漳　州	Zhangzhou	628.50	154.80	255.20	218.50
宁　德	Ningde	343.60	83.90	119.00	140.70

续表 continued

沿海城市 Coastal City		地区生产总值 Gross Regional Product	第一产业 Primary Industry	第二产业 Secondary Industry	第三产业 Tertiary Industry
山　东	**Shandong**	**9 609.20**	**895.00**	**5 721.50**	**2 992.40**
青　岛	Qingdao	2 695.80	178.30	1 396.30	1 121.20
东　营	Dongying	1 166.10	48.20	959.40	158.50
烟　台	Yantai	2 012.50	197.40	1 198.00	617.00
潍　坊	Weifang	1 471.20	201.40	836.50	433.30
威　海	Weihai	1 169.80	109.10	725.40	335.20
日　照	Rizhao	426.50	69.40	204.20	152.80
滨　州	Binzhou	667.30	91.20	401.70	174.40
广　东	**Guangdong**	**18 742.80**	**978.10**	**9 162.60**	**8 323.50**
广　州	Guangzhou	5 154.20	130.20	2 045.20	2 978.80
深　圳	Shenzhen	4 950.90	9.70	2 633.40	2 307.70
珠　海	Zhuhai	635.00	19.30	339.20	276.40
汕　头	Shantou	651.40	44.50	332.30	274.50
江　门	Jiangmen	805.40	72.50	425.40	307.50
湛　江	Zhanjiang	658.10	145.00	296.80	216.30
茂　名	Maoming	808.20	196.90	306.60	304.80
惠　州	Huizhou	803.40	75.10	458.40	269.90
汕　尾	Shanwei	211.00	50.60	83.80	76.60
阳　江	Yangjiang	295.30	80.20	110.00	105.20
东　莞	Dongguan	2 181.60	19.60	1 236.20	925.80
中　山	Zhongshan	880.20	30.80	539.40	31.60
潮　州	Chaozhou	287.60	32.60	152.40	102.50
揭　阳	Jieyang	420.50	71.10	203.50	145.90
广　西	**Guangxi**	**481.90**	**146.60**	**169.70**	**165.60**
北　海	Beihai	181.60	45.70	68.60	67.30
防城港	Fangchenggang	94.80	22.60	36.00	36.10
钦　州	Qinzhou	205.50	78.30	65.10	62.20
海　南	**Hainan**	**386.00**	**48.00**	**106.20**	**231.80**
海　口	Haikou	311.80	26.50	87.30	198.00
三　亚	Sanya	74.20	21.50	18.90	33.80

注：本表各省数据为合计数。

Note: The data for the provinces are the totals.

10-6 沿海地带生产总值（2005年）
Gross Regional Product of Coastal Counties, 2005

单位：万元 Unit:10 000 yuan

沿海地带	Coastal County (District)	地区生产总值 Gross Regional Product	第一产业 Primary Industry	第二产业 Secondary Industry	第三产业 Tertiary Industry
合 计	**Total**	**140 066 028**	**23 057 861**	**69 774 412**	**43 759 405**
河 北	**Hebei**	**4 915 126**	**1 276 587**	**2 102 173**	**1 536 366**
滦 南	Luannan	1 356 454	288 308	679 267	388 879
乐 亭	Leting	1 150 433	410 429	377 721	362 283
唐 海	Tanghai	285 227	102 113	94 424	88 690
昌 黎	Changli	646 637	204 827	251 371	190 439
抚 宁	Funing	637 167	161 765	308 297	167 105
黄 骅	Huanghua	715 999	77 322	343 037	295 640
海 兴	Haixing	123 209	31 823	48 056	43 330
辽 宁	**Liaoning**	**10 615 157**	**2 601 062**	**5 296 721**	**2 717 374**
长 海	Changhai	219 279	158 025	14 235	47 019
瓦房店	Wafangdian	2 339 745	404 745	1 385 000	550 000
普兰店	Pulandian	1 682 829	308 151	1 110 359	264 319
庄 河	Zhuanghe	1 648 258	382 800	939 619	325 839
东 港	Donggang	1 428 000	272 727	645 000	510 273
凌 海	Linghai	705 550	224 548	283 128	197 874
盖 州	Gaizhou	559 661	156 441	247 220	156 000
大 洼	Dawa	624 586	251 923	223 647	149 016
盘 山	Panshan	514 086	184 086	213 000	117 000
绥 中	Suizhong	621 406	185 298	162 827	273 281
兴 城	Xingcheng	271 757	72 318	72 686	126 753
江 苏	**Jiangsu**	**16 364 900**	**3 358 200**	**7 908 700**	**5 098 152**
海 安	Hai'an	1 489 000	239 900	759 700	489 400
如 东	Rudong	1 463 800	295 700	703 500	464 700

续表 continued

沿海地带 Coastal County (District)			地区生产总值 Gross Regional Product	第一产业 Primary Industry	第二产业 Secondary Industry	第三产业 Tertiary Industry
启	东	Qidong	2 000 300	332 000	1 029 900	638 400
通	州	Tongzhou	2 220 800	230 600	1 323 700	666 500
海	门	Haimen	2 126 100	221 600	1 247 900	656 600
东	海	Donghai	778 200	252 300	276 700	249 200
灌	云	Guanyun	464 800	169 400	157 600	137 800
灌	南	Guannan	400 500	131 400	144 000	125 100
响	水	Xiangshui	464 000	140 400	192 600	131 040
滨	海	Binhai	799 300	230 000	323 900	245 400
射	阳	Sheyang	1 217 400	366 000	482 300	369 112
东	台	Dongtai	1 673 100	411 500	735 700	525 900
大	丰	Dafeng	1 267 600	337 400	531 200	399 000
浙	**江**	**Zhejiang**	**37 241 207**	**3 084 826**	**21 255 951**	**12 900 430**
象	山	Xiangshan	1 363 760	269 855	602 308	491 597
宁	海	Ninghai	1 298 719	174 188	666 021	458 510
余	姚	Yuyao	2 981 160	225 316	1 751 293	1 004 551
慈	溪	Cixi	3 754 080	208 618	2 303 339	1 242 123
奉	化	Fenghua	1 248 225	122 521	639 696	486 008
洞	头	Dongtou	193 802	31 493	57 529	104 780
平	阳	Pingyang	1 057 863	82 367	518 884	456 612
苍	南	Cangnan	1 360 563	142 180	614 245	604 138
瑞	安	Rui'an	2 425 536	105 371	1 302 198	1 017 967
乐	清	Yueqing	2 601 892	115 977	1 611 056	874 859
海	盐	Haiyan	1 444 726	119 447	981 469	343 810
海	宁	Haining	2 179 587	142 242	1 327 804	709 541
平	湖	Pinghu	1 704 593	111 560	1 093 512	499 521
绍	兴	Shaoxing	3 868 200	169 100	2 506 900	1 192 200
上	虞	Shangyu	2 291 600	195 300	1 394 000	702 300
岱	山	Daishan	450 600	118 800	148 300	183 500

续表 continued

沿海地带 Coastal County (District)		地区生产总值 Gross Regional Product	第一产业 Primary Industry	第二产业 Secondary Industry	第三产业 Tertiary Industry
嵊　泗	Shengsi	338 600	66 100	153 300	119 200
玉　环	Yuhuan	1 486 704	130 510	912 521	443 673
三　门	Sanmen	491 658	103 391	191 697	196 570
温　岭	Wenling	3 051 154	286 396	1 618 477	1 146 281
临　海	Linhai	1 648 185	164 094	861 402	622 689
福　建	**Fujian**	**21 792 988**	**3 312 361**	**11 273 802**	**7 206 825**
连　江	Lianjiang	865 693	365 603	200 068	300 022
罗　源	Luoyuan	410 566	107 397	211 804	91 365
平　潭	Pingtan	401 882	148 523	61 269	192 090
福　清	Fuqing	2 639 631	397 141	1 414 743	827 747
长　乐	Changle	1 598 260	223 405	999 908	374 947
仙　游	Xianyou	683 769	131 281	279 458	273 030
惠　安	Hui'an	1 890 854	193 286	1 037 837	659 731
石　狮	Shishi	1 703 261	93 126	903 201	706 934
晋　江	Jinjiang	4 227 518	114 189	2 760 856	1 352 473
南　安	Nan'an	2 155 464	143 640	1 303 648	708 176
云　霄	Yunxiao	311 260	128 904	55 654	126 702
漳　浦	Zhangpu	711 551	266 969	153 930	290 652
诏　安	Zhao'an	442 273	182 898	110 929	148 446
东　山	Dongshan	326 103	125 059	101 866	99 178
龙　海	Longhai	1 553 802	244 189	958 685	350 928
霞　浦	Xiapu	511 240	164 074	129 082	218 084
福　安	Fu'an	759 798	137 303	353 196	269 299
福　鼎	Fuding	600 063	145 374	237 668	217 021
山　东	**Shandong**	**34 605 503**	**3 971 957**	**17 388 880**	**9 770 163**
胶　州	Jiaozhou	2 874 336	241 714	1 775 909	856 713
即　墨	Jimo	3 047 696	298 440	1 698 594	1 050 662

续表 continued

沿海地带 Coastal County (District)		地区生产总值 Gross Regional Product	第一产业 Primary Industry	第二产业 Secondary Industry	第三产业 Tertiary Industry
胶　南	Jiaonan	2 897 235	298 628	1 802 734	795 873
垦　利	Kenli	689 447	73 916	523 990	91 541
利　津	Lijin	487 478	117 340	266 204	103 934
广　饶	Guangrao	1 445 012	168 678	992 833	283 501
长　岛	Changdao	208 583	117 601	23 943	67 039
龙　口	Longkou	3 304 364	208 000	2 117 201	979 163
莱　阳	Laiyang	1 615 990	241 174	897 163	477 653
莱　州	Laizhou	2 339 691	313 569	1 346 322	679 800
蓬　莱	Penglai	1 701 556	151 475	1 069 344	480 737
招　远	Zhaoyuan	1 974 818	126 802	1 286 131	561 885
海　阳	Haiyang	1 006 962	249 045	465 425	292 492
昌　邑	Changyi	1 182 837	190 085	722 138	270 614
文　登	Wendeng	3 111 123	281 208	1 912 569	917 346
荣　成	Rongcheng	3 553 386	401 357	2 183 600	968 429
乳　山	Rushan	1 852 972	210 343	1 104 162	538 467
无　棣	Wudi	771 102	147 402	426 504	197 196
沾　化	Zhanhua	540 915	135 180	248 617	157 118
广　东	**Guangdong**	**10 311 457**	**3 296 948**	**3 645 462**	**3 369 047**
南　澳	Nan'ao	48 577	20 274	12 897	15 406
台　山	Taishan	1 094 855	200 782	524 979	369 094
恩　平	Enping	552 011	102 101	207 799	242 111
遂　溪	Suixi	573 912	301 414	109 426	163 072
徐　闻	Xuwen	405 978	198 083	55 626	152 269
廉　江	Lianjiang	771 332	332 352	211 161	227 819
雷　州	Leizhou	611 158	302 324	95 246	213 588
吴　川	Wuchuan	458 384	112 091	183 597	162 696
电　白	Dianbai	1 219 471	416 095	434 333	369 043

续表 continued

沿海地带 Coastal County (District)		地区生产总值 Gross Regional Product	第一产业 Primary Industry	第二产业 Secondary Industry	第三产业 Tertiary Industry
海　丰	Haifeng	776 103	148 097	301 615	326 391
陆　丰	Lufeng	643 713	213 182	230 992	199 539
阳　西	Yangxi	415 664	194 500	108 540	112 624
阳　东	Yangdong	541 127	161 583	231 121	148 423
饶　平	Raoping	721 262	209 943	250 510	260 809
揭　东	Jiedong	950 623	187 683	492 192	270 748
惠　来	Huilai	527 287	196 444	195 428	135 415
广　西	**Guangxi**	**136 041**	**41 905**	**34 091**	**60 045**
东　兴	Dongxing	136 041	41 905	34 091	60 045
海　南	**Hainan**	**4 083 649**	**2 114 015**	**868 632**	**1 101 003**
琼　海	Qionghai	512 452	292 057	57 854	162 541
儋　州	Danzhou	731 741	438 334	107 558	185 849
文　昌	Wenchang	540 525	274 021	98 254	168 250
万　宁	Wanning	412 240	165 896	76 541	169 803
东　方	Dongfang	396 671	128 995	181 185	86 491
澄　迈	Chengmai	406 500	158 672	164 410	83 418
临　高	Lingao	389 793	297 438	23 345	69 010
昌　江	Changjiang	264 115	82 640	132 848	48 628
乐　东	Ledong	244 961	160 089	15 104	69 768
陵　水	Lingshui	184 651	115 873	11 533	57 245

注：本表各省数据为合计数，不包含所有沿海城市市辖区数据，缺少上海崇明县，江苏赣榆县，福建金门县，山东寿光县，广东惠东县，广西合浦县数据。

Note: The data for the provinces are the totals, not including the data for all the municipal districts of the coastal cities; and lacking the data for Chongming of Shanghai Province; Ganyu of Jiangsu Province; Jinmen of Fujian Province; Shouguang of Shandong Province; Huidong of Guangdong and Hepu of Guangxi Province.

10-7 沿海地区财政收支（2006年）

Finencial Revenues and Expenditure by Coastal Regions, 2006

单位：万元 Unit:10 000 million yuan

地区 Region	财政收入 Government Revenue	#海域场地矿区使用费收入 Income from Use of Sea Area, Field and Diggings	财政支出 Government Expenditure	#基本建设 Expenditure for Capital Construction	#海域开发建设和场地使用费支出 Expenditure for Developing Land and Sea Area
合计 Total	**108 874 911**	**97 637**	**144 464 345**	**14 584 972**	**53 769**
天津 Tianjin	4 170 479	10 609	5 431 219	832 874	10 709
河北 Hebei	6 205 340	12 391	11 803 590	689 620	9 081
辽宁 Liaoning	8 176 718	16 939	14 227 471	1 372 691	4 297
上海 Shanghai	15 760 742		17 955 660	3 955 840	
江苏 Jiangsu	16 566 820	1 560	20 132 502	1 818 912	814
浙江 Zhejiang	12 982 044	3 260	14 718 593	1 001 210	621
福建 Fujian	5 411 707	11 223	7 286 973	555 755	5 384
山东 Shandong	13 562 526	20 375	18 334 400	821 963	10 695
广东 Guangdong	21 794 608	9 982	25 533 399	2 726 869	8 328
广西 Guangxi	3 425 788	6 420	7 295 172	621 530	2 925
海南 Hainan	818 139	4 878	1 745 366	187 708	915

注：1. 本表数据为预算执行数。

2. 本表数据为地方财政本级收入，不含财政部调整数。

Note: 1) Data in the table refer to budgetary data.

2) Data in the table refer to the own revenue of local governments which do not include the adjustment data of the Ministry of Finance.

10-8 沿海地区教育基本情况（2006年）
Basic Conditions on Education by Coastal Regions, 2006

地　区 Region	高等学校数 (所) Institutions of Higher Education (unit)	本、专科在校学生数 (人) Number of Students Enrolled in Undergraduate or Specialized Courses (person)	本、专科毕(结)业生数 (人) Number of Students Graduated in Undergraduate or Specialized Courses (person)
全国总计 **National Total**	**1 867**	**17 388 441**	**3 774 708**
天　津 Tianjin	45	357 382	81 983
河　北 Hebei	88	862 618	221 049
辽　宁 Liaoning	78	720 548	154 970
上　海 Shanghai	60	466 333	110 520
江　苏 Jiangsu	116	1 306 181	257 296
浙　江 Zhejiang	68	719 869	162 531
福　建 Fujian	63	461 341	94 979
山　东 Shandong	108	1 338 122	268 384
广　东 Guangdong	105	1 008 577	196 036
广　西 Guangxi	55	387 447	82 295
海　南 Hainan	15	90 138	16 436

10-9 沿海地区卫生基本情况（2006年）
Basic Conditions of Public Health by Coastal Regions, 2006

地　区 Region	卫生机构数 (个) Health Institutions (unit)	医疗机构床位数 (张) Total Beds (bed)	卫生机构人员 (人) Number of Employed Personnel in Health Institutions (person)
全国总计 **National Total**	**308 969**	**3 496 033**	**5 619 515**
天　津 Tianjin	2 367	43 600	78 631
河　北 Hebei	17 733	173 024	279 663
辽　宁 Liaoning	15 876	179 415	273 374
上　海 Shanghai	2 519	93 214	138 002
江　苏 Jiangsu	17 143	208 902	334 508
浙　江 Zhejiang	14 230	148 122	255 057
福　建 Fujian	9 652	84 536	124 909
山　东 Shandong	17 016	258 425	395 897
广　东 Guangdong	16 953	220 315	408 972
广　西 Guangxi	9 977	96 765	162 725
海　南 Hainan	2 337	19 927	38 199

10-10 沿海地区能源消耗指标（2006年）
Indicators of Energy Consumption by Coastal Regions, 2006

地 区 Region	单位地区生产总值能耗（等价值）Energy Consumption per Unit of GRP (equivalent value)		单位工业增加值能耗（规模以上，当量值）Energy Consumption per Unit of Industrial Value-added (above Designated Size,equivalent weight)	
	指标值（吨标准煤/万元）Index (ton of SCE/10 000 yuan)	上升或下降（±%）Change（±%）	指标值（吨标准煤/万元）Index (ton of SCE/10 000 yuan)	上升或下降（±%）Change（±%）
天 津 Tianjin	1.069	-3.98	1.33	-7.99
河 北 Hebei	1.895	-3.09	4.19	-5.59
辽 宁 Liaoning	1.775	-3.20	2.92	-6.10
上 海 Shanghai	0.873	-3.71	1.20	-6.00
江 苏 Jiangsu	0.891	-3.50	1.57	-7.32
浙 江 Zhejiang	0.864	-3.52	1.43	-1.90
福 建 Fujian	0.907	-3.20	1.37	-5.30
山 东 Shandong	1.231	-3.46	2.02	-5.70
广 东 Guangdong	0.771	-2.93	1.04	-2.96
广 西 Guangxi	1.191	-2.50	2.88	-7.36
海 南 Hainan	0.905	-1.17	3.15	-5.56

续表continued

地　区 Region	单位地区生产总值电耗（等价值） Electricity Consumption per Unit of GRP (equivalent value)	
	指标值 (千瓦小时/万元) Index (kw.h/10 000 yuan)	上升或下降 （±%） Change （±%）
天　津 Tianjin	1 024.1	-1.61
河　北 Hebei	1 515.9	1.90
辽　宁 Liaoning	1 372.6	-1.01
上　海 Shanghai	964.3	-4.15
江　苏 Jiangsu	1 221.2	1.92
浙　江 Zhejiang	1 247.4	2.06
福　建 Fujian	1 149.8	-0.17
山　东 Shandong	1 069.0	3.55
广　东 Guangdong	1 172.1	-1.95
广　西 Guangxi	1 252.0	0.02
海　南 Hainan	970.3	6.36

注：地区生产总值和工业增加值按2005年价格计算。

Note：Gross regional product and industrial value-added are caculated at 2005 constant prices.

10-11 沿海地区用水情况（2006）
Water Use of Coastal Regions, 2006

地　区 Region	用水总量 （亿立方米） Water Use (100 million m^3)	人均用水量 （立方米/人） Per Capita Water Use (m^3/person)
全国总计 National Total	**5 795.0**	**442.0**
天　津 Tianjin	23.0	216.8
河　北 Hebei	204.0	296.7
辽　宁 Liaoning	141.2	332.6
上　海 Shanghai	118.6	660.0
江　苏 Jiangsu	546.4	727.3
浙　江 Zhejiang	208.3	421.7
福　建 Fujian	187.3	528.0
山　东 Shandong	225.8	243.4
广　东 Guangdong	459.4	496.7
广　西 Guangxi	314.4	670.5
海　南 Hainan	46.5	558.4

10-12 沿海地区全社会固定资产投资（2006年）

Total Investment in Fixed Assets by the Whole Society of Coastal Regions, 2006

单位：亿元 Unit:100 million yuan

地 区 Region	全社会固定资产投资 Total Investment in Fixed Assets in the Whole Country	#农、林、牧、渔业 Agriculture,Forestry, Animal Husbandry and Fishery	#交通运输、仓储和邮政业 Transport,Storage and Post
全国总计 National Total	**109 998.2**	**2 749.9**	**12 138.1**
天 津 Tianjin	1 820.5	14.6	179.4
河 北 Hebei	5 470.2	239.2	548.9
辽 宁 Liaoning	5 689.6	155.5	571.7
上 海 Shanghai	3 900.0	14.3	600.8
江 苏 Jiangsu	10 069.2	65.4	632.3
浙 江 Zhejiang	7 590.2	61.2	865.0
福 建 Fujian	2 981.8	50.7	375.4
山 东 Shandong	11 111.4	315.2	544.3
广 东 Guangdong	7 973.4	60.7	807.7
广 西 Guangxi	2 198.7	80.8	229.4
海 南 Hainan	423.9	25.6	76.2

10-13 沿海地区按经营单位所在地分货物进出口总额（2006年）

Total Value of Imports and Exports by Location of Importers/Exporters of Coastal Regions, 2006

单位：万美元 Unit:10 000 US$

地　区 Region	进出口 Total	出口 Exports	进口 Imports
全国总计 National Total	**176 039 647**	**96 893 560**	**79 146 087**
天　津 Tianjin	6 446 194	3 349 078	3 097 116
河　北 Hebei	1 853 088	1 283 400	569 688
辽　宁 Liaoning	4 839 024	2 831 942	2 007 083
上　海 Shanghai	22 752 420	11 358 927	11 393 493
江　苏 Jiangsu	28 397 838	16 040 962	12 356 876
浙　江 Zhejiang	13 914 161	10 089 056	3 825 105
福　建 Fujian	6 265 963	4 126 174	2 139 789
山　东 Shandong	9 521 381	5 859 834	3 661 548
广　东 Guangdong	52 719 910	30 194 643	22 525 267
广　西 Guangxi	666 756	359 297	307 459
海　南 Hainan	284 621	137 562	147 059

10-14 沿海地区城镇居民平均每人全年家庭收入和消费性支出（2006年）

Per Capita Annual Income and Consumption Expenditure of Urban Households by Coastal Regions, 2006

单位：元 Unit:yuan

地　区 Region	总收入 Total Income	可支配收入 Disposable Income	消费性支出 Consumption Expenditure
全国平均水平 National Average	**12 719.19**	**11 759.45**	**8 696.55**
天　津 Tianjin	15 476.04	14 283.09	10 548.05
河　北 Hebei	10 887.19	10 304.56	7 343.49
辽　宁 Liaoning	11 230.03	10 369.61	7 987.49
上　海 Shanghai	22 808.57	20 667.91	14 761.75
江　苏 Jiangsu	15 248.66	14 084.26	9 628.59
浙　江 Zhejiang	19 954.03	18 265.10	13 348.51
福　建 Fujian	15 102.39	13 753.28	9 807.71
山　东 Shandong	13 222.85	12 192.24	8 468.40
广　东 Guangdong	17 725.56	16 015.58	12 432.22
广　西 Guangxi	10 624.30	9 898.75	6 791.95
海　南 Hainan	10 081.70	9 395.13	7 126.78

10-15 沿海地区农村居民家庭人均纯收入和生活消费支出（2006年）

Per Capita Annual Net Income and Consumption Expenditure of Rural Households by Coastal Regions, 2006

单位：元 Unit:yuan

地 区 Region	纯收入 Net Income	生活消费支出 Consumption Expenditure
全国平均水平 National Average	**3 587.04**	**2 829.02**
天 津 Tianjin	6 227.94	3 341.06
河 北 Hebei	3 801.82	2 495.33
辽 宁 Liaoning	4 090.40	3 066.87
上 海 Shanghai	9 138.65	8 006.00
江 苏 Jiangsu	5 813.23	4 135.21
浙 江 Zhejiang	7 334.81	6 057.16
福 建 Fujian	4 834.75	3 591.40
山 东 Shandong	4 368.33	3 143.80
广 东 Guangdong	5 079.78	3 885.97
广 西 Guangxi	2 770.48	2 413.93
海 南 Hainan	3 255.53	2 232.19

10-16 沿海地区人口情况（2006年）
Population of Coastal Regions, 2006

单位：万人　Unit:10 000 persons

地　区 Region	年末总人口 Total Population by the End of the Year	城镇人口 Urban Population
全国总计 National Total	**131 448**	**57 706**
天　津 Tianjin	1 075	814
河　北 Hebei	6 898	2 652
辽　宁 Liaoning	4 271	2 519
上　海 Shanghai	1 815	1 610
江　苏 Jiangsu	7 550	3 918
浙　江 Zhejiang	4 980	2 814
福　建 Fujian	3 558	1 708
山　东 Shandong	9 309	4 291
广　东 Guangdong	9 304	5 862
广　西 Guangxi	4 719	1 635
海　南 Hainan	836	385

注：本表数据根据2006年人口变动情况抽样调查数据推算。

Note:Data in the table are estimates from the 2006 National Sample Survey on Population Changes.

10-17 沿海城市人口情况（2005年）
Population of Coastal Cities, 2005

单位：万人 Unit:10 000 persons

沿海城市	Coastal City	年末总人口 Total Population by the End of the Year
合 计	**Total**	**23 403.5**
天 津	**Tianjin**	**1 043.0**
河 北	**Hebei**	**1 698.8**
唐 山	Tangshan	726.5
秦皇岛	Qinhuangdao	287.7
沧 州	Cangzhou	684.6
辽 宁	**Liaoning**	**1 746.1**
大 连	Dalian	565.3
丹 东	Dandong	242.4
锦 州	Jinzhou	308.3
营 口	Yingkou	230.5
盘 锦	Panjin	125.9
葫芦岛	Huludao	273.7
上 海	**Shanghai**	**1 360.3**
江 苏	**Jiangsu**	**2 041.8**
南 通	Nantong	770.9
连云港	Lianyungang	472.2
盐 城	Yancheng	798.7

续表 continued

沿海城市 Coastal City		年末总人口 Total Population by the End of the Year
浙　江	**Zhejiang**	**3 393.5**
杭　州	Hangzhou	660.5
宁　波	Ningbo	556.7
温　州	Wenzhou	750.3
嘉　兴	Jiaxing	334.3
绍　兴	Shaoxing	435.1
舟　山	Zhoushan	96.7
台　州	Taizhou	559.9
福　建	**Fujian**	**2 524.9**
福　州	Fuzhou	614.8
厦　门	Xiamen	153.2
莆　田	Putian	304.8
泉　州	Quanzhou	667.7
漳　州	Zhangzhou	457.4
宁　德	Ningde	327.0
山　东	**Shandong**	**3 489.5**
青　岛	Qingdao	819.6
东　营	Dongying	194.8
烟　台	Yantai	693.5
潍　坊	Weifang	871.6
威　海	Weihai	277.6
日　照	Rizhao	269.5
滨　州	Binzhou	362.9

续表 continued

沿海城市	Coastal City	年末总人口 Total Population by the End of the Year
广 东	**Guangdong**	**5 337.0**
广 州	Guangzhou	750. 5
深 圳	Shenzhen	181. 9
珠 海	Zhuhai	89. 6
汕 头	Shantou	491. 3
江 门	Jiangmen	386. 2
湛 江	Zhanjiang	714. 8
茂 名	Maoming	679. 3
惠 州	Huizhou	297. 6
汕 尾	Shanwei	315. 4
阳 江	Yangjiang	264. 2
东 莞	Dongguan	165. 7
中 山	Zhongshan	140. 8
潮 州	Chaozhou	250. 4
揭 阳	Jieyang	609. 3
广 西	**Guangxi**	**570.1**
北 海	Beihai	149. 2
防城港	Fangchenggang	79. 8
钦 州	Qinzhou	341. 1
海 南	**Hainan**	**198.5**
海 口	Haikou	147. 3
三 亚	Sanya	51. 2

注：本表各省数据为合计数。

Note: The data for the provinces are the totals.

10-18 沿海地带人口情况(2005年)

Population of Coastal Counties, 2005

单位：人　Unit: person

沿海地带	Coastal County (District)	年末总人口 Total Population by the End of the Year
合　计	**Total**	**80 661 814**
河　北	**Hebei**	**2 986 465**
滦　南	Luannan	577 089
乐　亭	Leting	492 845
唐　海	Tanghai	137 480
昌　黎	Changli	549 636
抚　宁	Funing	516 268
黄　骅	Huanghua	497 217
海　兴	Haixing	215 930
辽　宁	**Liaoning**	**6 686 331**
长　海	Changhai	74 874
瓦房店	Wafangdian	1 025 706
普兰店	Pulandian	825 036
庄　河	Zhuanghe	916 562
东　港	Donggang	645 896
凌　海	Linghai	604 085
盖　州	Gaizhou	726 409
大　洼	Dawa	391 231
盘　山	Panshan	291 179
绥　中	Suizhong	628 465
兴　城	Xingcheng	556 888
江　苏	**Jiangsu**	**12 975 200**
海　安	Hai'an	953 300
如　东	Rudong	1 076 800
启　东	Qidong	1 128 700

续表 continued

沿海地带 Coastal County (District)		年末总人口 Total Population by the End of the Year
通　州	Tongzhou	1 262 500
海　门	Haimen	1 013 600
东　海	Donghai	1 152 600
灌　云	Guanyun	1 050 500
灌　南	Guannan	740 100
响　水	Xiangshui	577 600
滨　海	Binhai	1 090 400
射　阳	Sheyang	1 043 700
东　台	Dongtai	1 155 900
大　丰	Dafeng	729 500
浙　江	**Zhejiang**	**14 264 217**
象　山	Xiangshan	527 360
宁　海	Ninghai	585 476
余　姚	Yuyao	825 783
慈　溪	Cixi	1 015 410
奉　化	Fenghua	478 778
洞　头	Dongtou	123 352
平　阳	Pingyang	848 247
苍　南	Cangnan	1 231 086
瑞　安	Rui'an	1 134 801
乐　清	Yueqing	1 169 668
海　盐	Haiyan	364 302
海　宁	Haining	643 857
平　湖	Pinghu	483 121
绍　兴	Shaoxing	704 720
上　虞	Shangyu	773 690
岱　山	Daishan	195 019
嵊　泗	Shengsi	81 174
玉　环	Yuhuan	398 583
三　门	Sanmen	409 732

续表 continued

沿海地带 Coastal County (District)		年末总人口 Total Population by the End of the Year
温　岭	Wenling	1 150 903
临　海	Linhai	1 119 155
福　建	**Fujian**	**12 364 322**
连　江	Lianjiang	611 002
罗　源	Luoyuan	253 183
平　潭	Pingtan	382 617
福　清	Fuqing	1 214 937
长　乐	Changle	660 060
仙　游	Xianyou	1 010 495
惠　安	Hui'an	924 898
石　狮	Shishi	308 478
晋　江	Jinjiang	1 032 910
南　安	Nan'an	1 480 671
云　霄	Yunxiao	416 402
漳　浦	Zhangpu	813 209
诏　安	Zhao'an	571 881
东　山	Dongshan	203 319
龙　海	Longhai	786 731
霞　浦	Xiapu	517 336
福　安	Fu'an	611 095
福　鼎	Fuding	565 098
山　东	**Shandong**	**11 140 000**
胶　州	Jiaozhou	777 000
即　墨	Jimo	1 090 000
胶　南	Jiaonan	811 000
垦　利	Kenli	215 000
利　津	Lijin	293 000

续表 continued

沿海地带 Coastal County (District)		年末总人口 Total Population by the End of the Year
广　饶	Guangrao	483 000
长　岛	Changdao	43 000
龙　口	Longkou	630 000
莱　阳	Laiyang	877 000
莱　州	Laizhou	858 000
蓬　莱	Penglai	445 000
招　远	Zhaoyuan	563 000
海　阳	Haiyang	666 000
昌　邑	Changyi	678 000
文　登	Wendeng	643 000
荣　成	Rongcheng	665 000
乳　山	Rushan	577 000
无　棣	Wudi	440 000
沾　化	Zhanhua	386 000
广　东	**Guangdong**	**15 187 426**
南　澳	Nan'ao	72 600
台　山	Taishan	987 514
恩　平	Enping	496 367
遂　溪	Suixi	971 892
徐　闻	Xuwen	683 662
廉　江	Lianjiang	1 537 805
雷　州	Leizhou	1 501 098
吴　川	Wuchuan	1 019 767
电　白	Dianbai	1 261 317
海　丰	Haifeng	769 049
陆　丰	Lufeng	1 599 013
阳　西	Yangxi	475 469
阳　东	Yangdong	454 933

续表 continued

沿海地带 Coastal County (District)		年末总人口 Total Population by the End of the Year
饶　平	Raoping	971 747
揭　东	Jiedong	1 223 170
惠　来	Huilai	1 162 023
广　西	**Guangxi**	**109 000**
东　兴	Dongxing	109 000
海　南	**Hainan**	**4 948 853**
琼　海	Qionghai	463 409
儋　州	Danzhou	981 884
文　昌	Wenchang	539 739
万　宁	Wanning	566 808
东　方	Dongfang	392 989
澄　迈	Chengmai	508 357
临　高	Lingao	427 574
昌　江	Changjiang	241 299
乐　东	Ledong	489 574
陵　水	Lingshui	337 220

注：本表各省数据为合计数，不包含所有沿海城市市辖区数据，缺少上海崇明县，江苏赣榆县，福建金门县，山东寿光县，广东惠东县，广西合浦县数据。

Note: The data for the provinces are the totals,not including the data for all the municipal districts of the coastal cities; and lacking the data for Chongming of Shanghai Province; Ganyu of Jiangsu Province;Jinmen of Fujian Province; Shouguang of Shandong Province; Huidong of Guangdong and Hepu of Guangxi Province.

10-19 沿海地区就业人员情况（2006年）
Number of Employed Persons by Coastal Regions, 2006

单位：万人　Unit:10 000 persons

地　区 Region	就业人员 Number of Employed Persons	城镇就业人员 Urban Employed Persons
全国总计 National Total	**76 400.0**	**28 310.0**
天　津 Tianjin		252.6
河　北 Hebei		688.8
辽　宁 Liaoning		874.5
上　海 Shanghai		615.5
江　苏 Jiangsu		1 401.7
浙　江 Zhejiang		1 077.9
福　建 Fujian		609.6
山　东 Shandong		1 375.6
广　东 Guangdong		1 849.8
广　西 Guangxi		449.9
海　南 Hainan		125.2

10-20 沿海城市就业人员情况（2005年）
Number of Employed Persons by Coastal Cities, 2005

单位：万人　Unit:10 000 persons

沿海城市 Coastal City		就业人员 Number of Employed Persons	城镇就业人员 Urban Employed Persons
合　计	**Total**	**13 995.65**	**4 945.36**
天　津	**Tianjin**	**542.52**	**312.76**
河　北	**Hebei**	**913.32**	**244.24**
唐　山	Tangshan	388.17	114.92
秦皇岛	Qinhuangdao	158.60	53.74
沧　州	Cangzhou	366.55	75.58
辽　宁	**Liaoning**	**925.00**	**423.60**
大　连	Dalian	336.46	202.56
丹　东	Dandong	117.01	40.81
锦　州	Jinzhou	146.70	51.40
营　口	Yingkou	104.24	35.64
盘　锦	Panjin	93.08	56.28
葫芦岛	Huludao	127.51	36.91
上　海	**Shanghai**	**863.32**	**598.34**
江　苏	**Jiangsu**	**990.36**	**213.27**
南　通	Nantong	444.40	85.54
连云港	Lianyungang	227.60	42.04
盐　城	Yancheng	318.36	85.69
浙　江	**Zhejiang**	**2 334.64**	**728.33**
杭　州	Hangzhou	481.10	185.05
宁　波	Ningbo	415.10	131.40
温　州	Wenzhou	488.39	164.31
嘉　兴	Jiaxing	240.83	87.18
绍　兴	Shaoxing	284.83	58.94
舟　山	Zhoushan	55.72	18.06
台　州	Taizhou	368.67	83.39
福　建	**Fujian**	**1 438.44**	**331.87**
福　州	Fuzhou	331.31	84.10
厦　门	Xiamen	139.42	70.31
莆　田	Putian	158.56	27.97

续表 continued

沿海城市 Coastal City		就业人员 Number of Employed Persons	城镇就业人员 Urban Employed Persons
泉　州	Quanzhou	417.15	97.85
漳　州	Zhangzhou	240.71	36.12
宁　德	Ningde	151.29	15.52
山　东	**Shandong**	**1 983.60**	**610.60**
青　岛	Qingdao	471.00	198.90
东　营	Dongying	109.80	49.40
烟　台	Yantai	406.20	132.80
潍　坊	Weifang	438.40	103.80
威　海	Weihai	151.90	57.70
日　照	Rizhao	183.20	25.90
滨　州	Binzhou	223.10	42.10
广　东	**Guangdong**	**3 633.74**	**1 370.51**
广　州	Guangzhou	574.46	334.90
深　圳	Shenzhen	576.30	377.58
珠　海	Zhuhai	94.01	84.61
汕　头	Shantou	116.60	30.22
江　门	Jiangmen	214.53	73.30
湛　江	Zhanjiang	305.00	53.55
茂　名	Maoming	308.10	42.57
惠　州	Huizhou	222.62	94.43
汕　尾	Shanwei	117.77	47.90
阳　江	Yangjiang	146.63	29.39
东　莞	Dongguan	388.13	40.96
中　山	Zhongshan	188.85	106.07
潮　州	Chaozhou	129.18	27.60
揭　阳	Jieyang	251.56	27.43
广　西	**Guangxi**	**256.91**	**44.51**
北　海	Beihai	7.20	8.71
防城港	Fangchenggang	44.84	9.02
钦　州	Qinzhou	204.87	26.78
海　南	**Hainan**	**113.80**	**67.33**
海　口	Haikou	90.40	59.16
三　亚	Sanya	23.40	8.17

注：本表各省数据为合计数。

Note: The data for the provinces are the totals.

10-21 沿海地带就业人员情况(2005年)

Number of Employed Persons of Coastal Counties, 2005

单位：人　Unit:person

沿海地带 Coastal County (District)	就业人员 Number of Employed Persons	城镇就业人员 Urban Employed Persons
合　计　Total	**41 760 372**	**5 654 870**
河　北　Hebei	**1 560 250**	**187 272**
滦　南　Luannan	307 937	28 890
乐　亭　Leting	279 789	21 260
唐　海　Tanghai	113 161	45 719
昌　黎　Changli	295 975	21 382
抚　宁　Funing	270 138	30 280
黄　骅　Huanghua	185 523	29 293
海　兴　Haixing	107 727	10 448
辽　宁　Liaoning	**3 060 370**	**445 138**
长　海　Changhai	43 934	6 777
瓦房店　Wafangdian	408 541	47 739
普兰店　Pulandian	356 347	53 174
庄　河　Zhuanghe	369 336	42 180
东　港　Donggang	302 328	24 655
凌　海　Linghai	253 487	20 996
盖　州　Gaizhou	317 725	15 892
大　洼　Dawa	311 666	139 745
盘　山　Panshan	214 206	61 480
绥　中　Suizhong	266 105	15 385
兴　城　Xingcheng	216 695	17 115
江　苏　Jiangsu	**6 228 356**	**629 956**
海　安　Hai'an	485 548	58 048

续表 continued

沿海地带 Coastal County (District)	就业人员 Number of Employed Persons	城镇就业人员 Urban Employed Persons
如　东 Rudong	588 412	53 012
启　东 Qidong	702 004	59 104
通　州 Tongzhou	646 402	56 602
海　门 Haimen	596 231	59 031
东　海 Donghai	530 094	40 494
灌　云 Guanyun	424 018	41 818
灌　南 Guannan	335 475	25 775
响　水 Xiangshui	226 026	33 626
滨　海 Binhai	409 446	33 446
射　阳 Sheyang	398 007	62 707
东　台 Dongtai	525 312	56 912
大　丰 Dafeng	361 381	49 381
浙　江 Zhejiang	**9 233 800**	**1 347 300**
象　山 Xiangshan	458 900	172 700
宁　海 Ninghai	377 500	20 800
余　姚 Yuyao	501 200	58 400
慈　溪 Cixi	816 500	52 300
奉　化 Fenghua	288 700	30 200
洞　头 Dongtou	61 400	7 400
平　阳 Pingyang	511 000	73 700
苍　南 Cangnan	732 700	67 200
瑞　安 Rui'an	703 700	89 100
乐　清 Yueqing	851 800	185 900
海　盐 Haiyan	228 800	56 300
海　宁 Haining	417 700	104 600
平　湖 Pinghu	314 500	99 200
绍　兴 Shaoxing	418 100	40 900
上　虞 Shangyu	471 200	87 100
岱　山 Daishan	95 000	9 900

续表 continued

沿海地带 Coastal County (District)	就业人员 Number of Employed Persons	城镇就业人员 Urban Employed Persons
嵊　泗 Shengsi	40 900	8 500
玉　环 Yuhuan	303 500	27 700
三　门 Sanmen	245 100	24 400
温　岭 Wenling	739 800	68 100
临　海 Linhai	655 800	62 900
福　建 Fujian	**6 683 461**	**1 115 211**
连　江 Lianjiang	329 964	39 083
罗　源 Luoyuan	99 930	12 030
平　潭 Pingtan	192 353	15 266
福　清 Fuqing	689 854	157 839
长　乐 Changle	301 513	42 505
仙　游 Xianyou	513 222	46 958
惠　安 Hui'an	557 667	109 992
石　狮 Shishi	182 336	66 230
晋　江 Jinjiang	771 201	290 971
南　安 Nan'an	873 256	82 605
云　霄 Yunxiao	185 779	20 409
漳　浦 Zhangpu	446 166	51 193
诏　安 Zhao'an	306 457	21 011
东　山 Dongshan	94 380	16 958
龙　海 Longhai	420 099	74 303
霞　浦 Xiapu	230 496	18 044
福　安 Fu'an	210 329	32 021
福　鼎 Fuding	278 459	17 793
山　东 Shandong	**5 996 717**	**1 126 697**
胶　州 Jiaozhou	470 084	153 740
即　墨 Jimo	687 418	130 518

续表 continued

沿海地带 Coastal County (District)	就业人员 Number of Employed Persons	城镇就业人员 Urban Employed Persons
胶　南 Jiaonan	406 947	71 607
垦　利 Kenli	83 796	20 296
利　津 Lijin	164 582	16 782
广　饶 Guangrao	251 441	27 741
长　岛 Changdao	18 077	4 977
龙　口 Longkou	324 294	71 994
莱　阳 Laiyang	431 382	67 282
莱　州 Laizhou	473 057	80 603
蓬　莱 Penglai	245 300	35 400
招　远 Zhaoyuan	282 223	78 277
海　阳 Haiyang	353 826	35 300
昌　邑 Changyi	318 849	32 129
文　登 Wendeng	369 637	67 137
荣　成 Rongcheng	265 323	66 423
乳　山 Rushan	321 291	48 491
无　棣 Wudi	292 290	66 000
沾　化 Zhanhua	236 900	52 000
广　东 Guangdong	**7 047 802**	**591 030**
南　澳 Nan'ao	30 035	5 635
台　山 Taishan	527 188	45 988
恩　平 Enping	217 176	30 676
遂　溪 Suixi	454 788	41 788
徐　闻 Xuwen	328 765	37 684
廉　江 Lianjiang	733 001	53 001
雷　州 Leizhou	627 127	58 027
吴　川 Wuchuan	519 908	33 908
电　白 Dianbai	635 527	50 727
海　丰 Haifeng	369 400	29 600

续表 continued

沿海地带 Coastal County (District)	就业人员 Number of Employed Persons	城镇就业人员 Urban Employed Persons
陆丰 Lufeng	609 467	45 567
阳西 Yangxi	279 940	21 640
阳东 Yangdong	269 899	26 808
饶平 Raoping	435 968	42 968
揭东 Jiedong	612 286	29 086
惠来 Huilai	397 327	37 927
广西 Guangxi	**55 146**	**5 246**
东兴 Dongxing	55 146	5 246
海南 Hainan	**1 933 070**	**207 020**
琼海 Qionghai	187 173	18 512
儋州 Danzhou	313 284	31 307
文昌 Wenchang	221 208	24 464
万宁 Wanning	180 888	22 301
东方 Dongfang	187 585	19 945
澄迈 Chengmai	196 823	29 708
临高 Lingao	183 399	14 022
昌江 Changjiang	97 726	19 426
乐东 Ledong	224 089	17 026
陵水 Lingshui	140 895	10 309

注：本表各省数据为合计数，不包含所有沿海城市市辖区数据，缺少上海崇明县，江苏赣榆县，福建金门县，山东寿光县，广东惠东县，广西合浦县数据。

Note: The data for the provinces are the totals,not including the data for all the municipal districts of the coastal cities;and lacking the data for Chongming of Shanghai Province; Ganyu of Jiangsu Province;Jinmen of Fujian Province; Shouguang of Shandong Province; Huidong of Guangdong and Hepu of Guangxi Province.

主要统计指标解释

1.国内(或地区)生产总值 指一个国家(或地区)所有常驻单位在一定时期内生产活动的最终成果。国内生产总值有三种表现形态,即价值形态、收入形态和产品形态。从价值形态看,它是所有常驻单位在一定时期内生产的全部货物和服务价值超过同期中间投入的全部非固定资产货物和服务价值的差额,即所有常驻单位的增加值之和;从收入形态看,它是所有常驻单位在一定时期内创造并分配给常驻单位和非常驻单位的初次收入分配之和;从产品形态看,它是所有常驻单位在一定时期内最终使用的货物和服务价值与货物和服务净出口价值之和。在实际核算中,国内生产总值有三种计算方法,即生产法、收入法和支出法。三种方法分别从不同的方面反映国内生产总值及其构成。

2.三次产业 是根据社会生产活动历史发展的顺序对产业结构的划分,产品直接取自自然界的部门称为第一产业,对初级产品进行再加工的部门称为第二产业,为生产和消费提供各种服务的部门称为第三产业。它是世界上较为通用的产业结构分类,但各国的划分不尽一致。我国的三次产业划分是:

第一产业:是指农、林、牧、渔业。

第二产业:是指采矿业,制造业,电力、燃气及水的生产和供应业,建筑业。

第三产业:是指除第一、二产业以外的其他行业。第三产业包括:交通运输、仓储和邮政业,信息传输、计算机服务和软件业,批发和零售业,住宿和餐饮业,金融业,房地产业,租赁和商务服务业,科学研究、技术服务和地质勘查业,水利、环境和公共设施管理业,居民服务和其他服务业,教育,卫生、社会保障和社会福利业,文化、体育和娱乐业,公共管理和社会组织,国际组织。

3.增加值 是指各行各业生产经营和劳务活动的最终成果,采用生产法和收入法两种方法计算。

生产法 是从货物和服务活动在生产过程中形成的总产品入手,剔除生产过程中投入的中间产品价值,得到新增价值的方法。

收入法 又称分配法。按收入法计算国内生产总值是从生产过程创造的收入的角度对常驻单位的生产活动成果进行核算;按照此法计算,增加值由劳动者报酬、固定资产折旧、生产税净额和营业盈余四个部分组成。

4.年末总人口 是指每年12月31日24时一定地区范围内的有生命的个人的人口总和。

5.财政收入 指国家财政参与社会产品分配所取得的收入,是实现国家职能的财力保证。财政收入所包括的内容几经变化,目前主要包括:

(1)各项税收:包括增值税、营业税、消费税、土地增值税、城市维护建设税、资源税、城市土地使用税、企业所得税、个人所得税、关税、证券交易印花税、车辆购置税、农牧业税和耕地占用税等。

(2)专项收入:包括排污费收入、城市水资源费收入、矿产资源补偿费收入、教育费附加收入等。

(3)其他收入:包括利息收入、基本建设贷款归还收入、基本建设收入、捐赠收入等。

(4)国有企业亏损补贴:此项为负收入,冲减财政收入。主要包括对工业企业、商业企业、粮食企业的补贴。

6. 财政支出 国家财政将筹集起来的资金进行分配使用,以满足经济建设和各项事业的需要,

7.基本建设支出 指按国家有关规定,属于基本建设范围内的基本建设有偿使用、拨款、资本金支出以及经国家批准对专项和政策性基建投资贷款,在部门的基建投资额中统筹支付的贴息支出。

8.普通高等学校 指按照国家规定的设置标准和审批程序批准举办的,通过全国普通高等学校统一招生考试,招收高中毕业生为主要培养对象,实施高等教育的全日制大学、独立设置的学院和高等专科学校、高等职业学校和其他机构。

大学、独立设置的学院主要实施本科层次以上教育,高等专科学校、高等职业学校实施专科层次教育,其他机构是承担国家普通招生计划任务不计校数的机构。包括普通高等学校分校和批准筹建的普通高等学校等。

9.卫生机构 包括医疗机构、疾病预防控制中心(防疫站)、采供血机构、卫生监督及监测(检验)机构、医学科研和在职培训机构、健康教育所等。

10.医疗机构 包括医院、社区卫生服务中心(站)、疗养院、卫生院、门诊部、诊所(卫生所、医务室)、妇幼保健院(所、站)、专科疾病防治院(所、站)、急救中心(站)和临床检验中心。医疗机构分为非赢利性医疗机构和赢利性医疗机构。

11.单位国内生产总值能耗 指一定时期内,一个国家或地区每生产一个单位的国内生产总值所消耗的能源。计算公式为:

$$单位国内生产总值能源=\frac{能源消费总量}{国内生产总值}$$

12.单位国内生产总值电耗 指一定时期内,一个国家或地区每生产一个单位的国内生产总值所消耗的电力。计算公式为:

$$单位国内生产总值电耗=\frac{全社会用电量}{国内生产总值}$$

13.单位工业增加值能耗 指一定时期内,一个国家或地区每生产一个单位的工业增加值所消耗的能源。计算公式为:

$$单位工业增加值能耗=\frac{工业能源消费总量}{工业增加值}$$

14.用水总量 指分配给各类用户的包括输水损失在内的毛用水量之和,不包括海水直接利用量。

15.全社会固定资产投资 是以货币形式表现的在一定时期内全社会建造和购置固定资产的工作量以及与此有关的费用的总称。该指标是反映固定资产投资规模、结构和发展速度的综合性指标,又是观察工程进度和考核投资效果的重要依据。全社会固定资产投资按登记注册类型可分为国有、集体、个体、联营、股份制、外商、港澳台商、其他等。

16.进出口总额 指实际进出我国国境的货物总金额。包括对外贸易实际进出口货物,来料加工装配进出口货物,国家间、联合国及国际组织无偿援助物资和赠送品,华侨、港澳台同胞和外籍华人捐赠品,租赁期满归承租人所有的租赁货物,进料加工进出口货物,边境地方贸易及边境地区小额贸易进出口货物(边民互市贸易除外),中外合资企业、中外合作经营企业、外

商独资经营企业进出口货物和公用物品,到、离岸价格在规定限额以上的进出口货样和广告品(无商业价值、无使用价值和免费提供出口的除外),从保税仓库提取在中国境内销售的进口货物,以及其他进出口货物。该指标可以观察一个国家在对外贸易方面的总规模。我国规定出口货物按离岸价格统计,进口货物按到岸价格统计。

17.商品经营单位所在地进、出口额 指在所在地海关注册登记的有进出口经营权的企业实际进、出口额。

18.城镇家庭可支配收入 指家庭成员得到可用于最终消费支出和其它非义务性支出以及储蓄的总和,即居民家庭可以用来自由支配的收入。它是家庭总收入扣除交纳的所得税、个人交纳的社会保障支出以及记账补贴后的收入。计算公式为:

可支配收入=家庭总收入-交纳所得税-个人交纳的社会保障支出-记账补贴

19.城镇家庭消费性支出 指家庭用于日常生活的支出,包括食品、衣着、家庭设备用品及服务、医疗保健、交通和通信、娱乐教育文化服务、居住、杂项商品和服务等八大类支出。

20.总收入 指调查期内农村住户和住户成员从各种来源渠道得到的收入总和。按收入的性质划分为工资性收入、家庭经营收入、财产性收入和转移性收入。

21.纯收入 指农村住户当年从各个来源得到的总收入相应地扣除所发生的费用后的收入总和。计算方法:

纯收入=总收入-税费支出-家庭经营费用支出-生产性固定资产折旧-赠送农村亲友支出

纯收入主要用于再生产投入和当年生活消费支出,也可用于储蓄和各种非义务性支出。"农民人均纯收入"按人口平均的纯收入水平,反映的是一个地区或一个农户农村居民的平均收入水平。

22.人口数 指一定时点、一定地区范围内有生命的个人总和。

年度统计的年末人口数指每年 12 月 31 日 24 时的人口数。年度统计的全国人口总数内未包括香港、澳门特别行政区和台湾省以及海外华侨人数。

23.城镇人口 城镇人口是指居住在城镇范围内的全部常住人口

24.就业人员 指在 16 周岁及以上,从事一定社会劳动并取得劳动报酬或经营收入的人员。这一指标反映了一定时期内全部劳动力资源的实际利用情况,是研究我国基本国情国力的重要指标。

Explanatory Notes on Main Statistical Indicators

1.Gross Domestic Product (GDP) refers to the final result of the primary distribution of the income created by all the resident units of a country (or a region) during a certain period of time. Gross domestic product is expressed in three different forms, i.e. value, income, and products respectively. The form of value refers to the total value of all products and services produced by all resident units during a certain period of time minus the total value of intermediate input of materials and services of the nature of non-fixed assets or the summation of the value added of all resident units: the form of income includes all the income created by all resident units and distributed primarily to all resident and non-resident units; the form of products refers to the summation of the value of the products and services finally used and the net export value of products and services by all resident units during a given period of time. In the practice of national accounting, gross domestic product is calculated with three approaches, i.e.production approach, income approach, and expenditure approach, which reflect the gross domestic product and its composition from different aspects.

2.Three Industries Industrial structure is classified according to the sequence of historical development of social productive activities. Primary industry refers to the extraction of natural resources; secondary industry involves processing of primary products; and tertiary industry provides services of various kinds for production and consumption. The above classification is universal in the world although it varies to some extent form country to country. The three industries in China are divided as follows:

Primary industry: refers to farming, forestry, sideline production and fishery.

Secondary industry: refers to such industries as mining, manufacturing, production and supply of electric power, fuel gas and water, and construction.

Tertiary industry: refers to all the other industries not included in the primary or secondary industries. It includes such industries as communications and transportation, storage and postal service; information transmission, computer service and software; wholesale and retailing; accommodation and catering; financial service; real estate; charter business and commercial affairs service; scientific research, technological service and geological survey; water conservancy, environmental and other public facilities management; residents service and other service trades; education, health, social security and social welfare; culture, sports and entertainment business as well as public administration and social organizations, and international organizations.

3.Added Value refers to the final result of production operation and labor activities of all trades and professions, which is calculated by using the methods of production and income.

Production Method: refers to the method whereby to get the newly added value by proceeding from the gross product of goods and service activities occurring in the course of production and then rejecting the value of intermediate product input in the course of production.

Income Method: is also called the distribution method. The calculation of the gross domestic product (GDP) by the income method is the accounting of the result of productive activities of permanent units from the angle of the income created in the course of production. According to this method, the added value is composed of payment for laborers, depreciation for fixed assets, net tax on production and business surplus.

4.Total Population by the End of the Year refers to the sum of living individuals within a particular range of area at 24:00 on December 31 of each year.

5.Government Revenue refers to the income obtained by the government finance through participating in the distribution of social products. It is the financial guarantee to ensure government functioning. The content of government revenue has changed several times. Now it includes the following main items:

(1) Various tax revenues, including value added tax, business tax, consumption tax, land value-added tax, tax on city maintenance and construction, resources tax, tax on use of urban land, enterprise income tax, personal income tax, tariff, stamp tax on security transactions, tax on purchase of motor vehicles, tax on agriculture and animal husbandry and tax on occupancy of cultivated land, etc.

(2) Special revenues, including revenues from the fee on sewage treatment, fee on urban water resources, fee for the compensation of mineral resources and extra-charges for education, etc.

(3) Other revenues, including revenues from interest, repayment of capital construction loan, capital construction projects, and donations and grants.

(4) Subsidies for the losses of State-owned enterprises. This is an item of negative revenue, counteracting revenues and consisting of subsidies to industrial, commercial and grain purchasing and supply enterprises.

6.Government Expenditure refers to the distribution and use of the funds which the government finance has raised, so as to meet the needs of economic construction and various causes.

7.Expenditure for capital construction It refers to the non-gratuitous use of, appropriation of funds for and capital outlay on capital construction in the area of capital construction. It also covers the loans on capital construction approved by the government for special purposes or policy purposes and the expenditure with discount paid in an overall way within the amount of the funds appropriated to the departments for capital construction.

8.Regular Institutions of Higher Learning refer to educational establishments set up according to the government evaluation and approval procedures, enrolling graduates from senior secondary schools and providing higher education courses and training for senior professionals. They include full-time universities, colleges, institutions of higher professional education, institutions of higher vocational education and others.

Universities and colleges primarily provide undergraduate courses; institutions of higher professional education and institutions of higher vocational education primarily provide professional trainings; and others refer to educational establishments, which are responsible for enrolling higher education students under the State Plan but not enumerated in the total number of schools, including: branch schools of universities and colleges, and universities and colleges that have been approved and under plan for construction.

9.Health Care Institutions include: medical institutions, disease prevention and control centres (epidemic prevention stations), blood gathering and supplying institutions, health supervision and inspection (check up) institutions, medicinal scientific research and on-job training institutions, health education centres and so on.

10.Medical Organizations include: hospitals, health service centres (stations) in communities, sanatoria, health centres, outpatient clinics, clinics (health stations and infirmaries), maternity and child care agencies (centres and stations), special disease prevention and curing agencies (centres and stations), first aid centres (stations) and clinical inspection centres. Medical organizations are grouped by two types: profit-making and non-profit-making medical organizations.

11.Energy Consumption per Unit of GDP refers to the energy consumption per unit of Gross Domestic Product in a country or the Gross Regional Product in a region in the same reference period. The formula is: $\text{Energy Consumption per Unit of GDP} = \frac{\text{Total Energy Consumtion}}{\text{Gross Domestic Product}}$

12.Electricity Consumption per Unit of GDP refers to the electricity consumption per unit of Gross Domestic Product in a country or the Gross Regional Product in a region in the same reference period. The formula is: $\text{Electricity Consumption per Unit of GDP} = \frac{\text{Total Electricity Consumtion}}{\text{Gross Domestic Product}}$

13.Energy Consumption per Unit of Industrial Value–added refers to the energy consumption per unit of industrial value-added in a country or region in the same reference period. The formula is: $\text{Energy Consumption per Unit of Industrial Value-added} = \frac{\text{Total Electricity Consumtion}}{\text{Gross Domestic Product}}$

14.Gross Amount of Water Used refers to the gross water use distributed to users, including loss during transportation, broken down into use by agriculture, industry, living consumption and ecological protection.

15.Total Investment in Fixed Assets in the Whole Country refers to the volume of activities in construction and purchases of fixed assets of the whole country and related fees, expressed in monetary terms during the reference period. It is a comprehensive indicator which shows the size, structure and growth of the investment in fixed assets, providing a basis for observing the progress of construction projects and evaluating the results of investment. Total investment in fixed assets in the whole country includes, by type of ownership, the investment by State-owned units, collective-owned units, individuals, joint ownership units, share-holding units, as well as investments by entrepreneurs from foreign countries and from Hong Kong, Macao and Taiwan, and by other units.

16.Total Imports and Exports at Customs refer to the real value of commodities imported and exported across the border of China. They include the actual imports and exports through foreign trade, imported and exported goods under the processing and assembling trades and materials, supplies and gifts as aid given gratis between governments and by the United Nations and other international organizations, and contributions donated by overseas Chinese compatriots in Hong Kong and Macao and Chinese with foreign citizenship, leasing commodities owned by tenant at the expiration of leasing period, the imported and exported commodities processed with imported materials, commodities trading in border areas (excluding mutual exchange goods), the imported and exported commodities and articles for public use of the Sino-foreign joint ventures, cooperative enterprises and ventures with sole foreign investment. Also included is the import or export of samples and advertising goods for which the CIF or FOB value is beyond the permitted ceiling (excluding goods of no trading or use value and free commodities for export), imported goods sold in China from bonded warehouses and other imported or exported goods. The indicator of the total imports and exports at customs can be used to observe the total size of external trade in a country. In accordance with the stipulation of the Chinese government, imports are calculated at CIF, while exports are calculated at FOB.

17.Import–Export Value by Location of China´s Foreign Trade Managing Units refers to actual value of imports and exports carried out by corporations which have been registered by the local customs house and are vested with right to run import export business.

18.Disposable Income of Urban Households refers to the actual income at the disposal of members of the households which can be used for final consumption, other non-compulsory expenditure and savings. This equals to total income minus income tax, personal contribution to social security and subsidy for keeping diaries in being a sampled household. The following formula is used:

Disposable income = total household income - income tax - personal contribution to social security - subsidy for keeping diaries for a sampled household

19.Consumption Expenditure of Urban Households refers to the total expenditure of households for consumption in daily life, including expenditure on the eight categories of food; clothing; household appliances and services; health care and medical services; transport and communications; recreation, education and cultural services; housing; and miscellaneous goods and services.

20.Total Income refers to the sum of income earned from various sources by the rural households and their members during the reference period, and is classified as income from wages and salaries, household operations, properties and transfers.

21.Net Income refers to the total income of rural households from all sources minus all corresponding expenses. The formula for calculation is as follows:

Net income = total income - taxes and fees paid - household operation expenses - taxes and fees - depreciation of fixed assets for production - gifts to non-rural relatives

Net income is mainly used as input for reinvestment in production and as consumption expenditure of the year, and also used for savings and non-compulsory expenses of various forms. "Per capita net income of farmers" is the level of net income averaged by population, reflecting the average income level of rural households in a given area.

22.Total Population refers to the total number of people alive at a certain point of time within a given area.

The annual statistics on total population is taken at midnight, the 3lst of December, not including residents in Taiwan province, Hong Kong and Macau, and overseas Chinese.

23.Urban Population refers to all people residing in cities and towns.

24.Employed Persons refers to persons aged 16 and over who are engaged in gainful employment and thus receive remuneration payment or earn business income. This indicator reflects the actual utilization of total labour force during a certain period of time and is often used for the research on China's economic situation and national power.

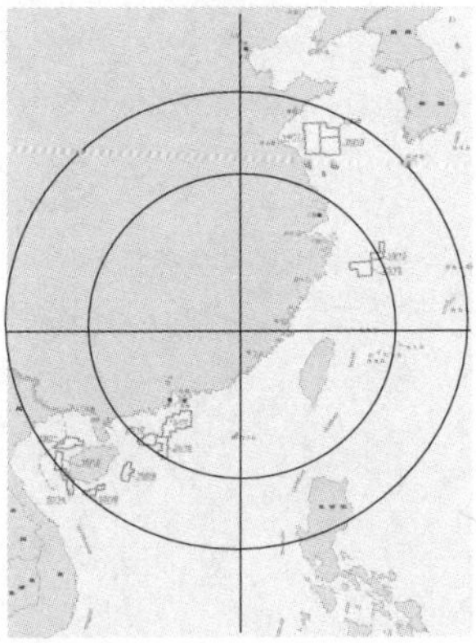

11

部分世界海洋经济统计资料

Part of the Worlds Marine Economic Statistics Data

11-1 世界海洋面积
The World Ocean Area

区 域 Region	海洋面积(平方公里) Ocean Area(km^2)	占世界海洋面积的比重（%） Proportion of The World Ocean Area (%)	占地球表面面积的比重（%） Proportion of The Earth's Surface Area (%)
合 计 Total	**361 000 000**	**100.0**	**70.77**
太 平 洋 Pacific Ocean	178 334 000	49.4	34.96
大 西 洋 Atlantic Ocean	91 694 000	25.4	17.98
印 度 洋 Indian Ocean	76 171 000	21.1	14.93
北 冰 洋 Arctic Ocean	14 801 000	4.1	2.90

资料来源：《2006/2007国际统计年鉴》（表11-3同）。

Source: *International Statistical Yearbook 2006/2007* (Same as Table 11-3).

11-2 主要沿海国家(地区)海岸线长度
Length of Coastline of Major Coastal Countries(Regions)

单位:公里　Unit:km

国家和地区 Country and Region	海岸线长度 Length of Coastline
美国 United States	22 680
日本 Japan	30 000
德国 Germany	1 300
英国 United Kingdom	11 450
法国 France	3 000
意大利 Italy	7 000
加拿大 Canada	20 000
澳大利亚 Australia	20 125
俄罗斯联邦 Russian Fed.	34 000
波兰 Poland	491
印度 India	6 083
印度尼西亚 Indonesia	35 000
菲律宾 Philippines	18 533
泰国 Thailand	3 219
马来西亚 Malaysia	4 675
新加坡 Singapore	193
缅甸 Myanmar	3 060
孟加拉国 Bangladesh	580
土耳其 Turkey	7 200
韩国 Korea Rep.	2 413
埃及 Egypt	2 450
墨西哥 Mexico	9 330
巴西 Brazil	7 400
阿根廷 Argentina	4 989

11-3 世界主要沿海国家（地区）土地和人口（2005年）
The Area of Territory and Population of Major Coastal Countries(Regions), 2005

国家和地区 Country and Region	国土面积(万平方公里) Area of Territory (10 000 km^2)	年中人口（万人） Mid-year Population (10 000 persons)	人口密度（人/平方公里） Population Density (people per km^2)
世　界 World	**13 427.9**	**643 778.0**	**47.9**
发达国家[①] Developed Countries	**5 643.3**	**101 133.7**	**17.9**
发展中国家[①] Developing Countries	**7 789.1**	**542 644.7**	**69.7**
中　国 China	960. 0	130 756[②]	136. 2
美　国 United States	962. 9	29 649. 7	30. 8
日　本 Japan	37. 8	12 795. 6	338. 6
加拿大 Canada	998. 5	3 227. 1	3. 2
德　国 Germany	35. 7	8 248. 5	231. 0
英　国 United Kingdom	24. 4	6 020. 3	247. 1
法　国 France	55. 2	6 074. 3	110. 1
意大利 Italy	30. 1	5 747. 1	190. 7
墨西哥 Mexico	195. 8	10 308. 9	52. 6
印度尼西亚 Indonesia	190. 5	22 055. 8	115. 8
马来西亚 Malaysia	33. 0	2 534. 7	76. 9
泰　国 Thailand	51. 3	6 423. 3	125. 2
新加坡 Singapore	0. 1	435. 1	4 351. 0
韩　国 Korea.Rep.	9. 9	4 829. 4	486. 5
印　度[③] India	328. 7	109 458. 3	333. 0
巴　西 Brazil	851. 5	18 640. 5	21. 9
俄罗斯联邦 Russian Federation	1 709. 8	14 315. 1	8. 4

注：①发达国家指高收入国家，发展中国家指中低收入国家。②中国为年底人口数。③为2003年数据。

Note:①Developed countries refer to those with high revenue,and developing counties are those with low and medium revenue.

②The data here refer to the population at the end of year for China.

③The data for 2003.

11-4 主要沿海国家(地区)国内生产总值

Gross Domestic Product of Major Coastal Countries(Regions)

单位:亿本币 Unit: National Currency 100 million

国家和地区 Country and Region	2005年			2006年
	国内生产总值 (亿本国币) GDP (National Currency 100 million)	人均国民总收入 (美元) GNI per Captia (current US$)	国内生产总值增长率(%) Growth Rate of GDP(%)	国内生产总值 (亿本国币) GDP (National Currency 100 million)
美 国 United States	124 857	43 740	3.5	132 446
日 本 Japan	5 029 054	38 980	2.7	5 078 096
德 国 Germany	22 455	34 580	0.9	23 072
英 国 United Kingdom	12 093	37 600	1.8	12 882 *
法 国 France	16 904	34 810	1.2	17 773
意大利 Italy	14 177	30 010		14 754
加拿大 Canada	13 687	32 600	2.9	14 393
俄罗斯联邦 Russian Fed.	216 650	4 460	6.4	266 213
印 度 India	341 953	720	8.5	395 007 *
印度尼西亚 Indonesia	26 786 641	1 280	5.6	33 381 957
泰 国 Thailand	69 243	2 750	4.5	78 131
马来西亚 Malaysia	4 945	4 960	5.2	5 463
新加坡 Singapore	1 944	27 490	6.4	2 100 *
韩 国 Korea Rep.	8 121 966	15 830	4	8 478 764
墨西哥 Mexico	83 743	7 310	3	91 555
巴 西 Brazil	19 303	3 460	2.3	23 228

注：*为估计数。

Note:* is Estimated data.

11-5 世界主要沿海国家（地区）国内生产总值产业构成(2005年)
The Industry Structure of GDP of Major Coastal Countries (Regions), 2005

国家和地区 Country and Region	国内生产总值产业构成（%） Industry Structure of GDP(%)		
	第一产业 Primary Industy	第二产业 Secondary Industry	第三产业 Tertiary Industy
世　界 World	**3.5**③	**28.0**③	**68.5**③
发达国家① **Developed Countries**	**1.6**③	**26.2**③	**72.2**③
发展中国家① **Developing Countries**	**11.7**④	**36.0**④	**52.3**④
中　国 China	12.6	47.5	39.9
美　国 United States	1.2③	22.3③	76.5③
日　本 Japan	1.3③	30.5③	68.2③
加拿大 Canada	2.2②	32.4②	65.4②
德　国 Germany	1.1④	29.1④	69.8④
英　国 United Kingdom	1.0④	26.3④	72.7④
法　国 France	2.5④	21.7④	75.7④
意大利 Italy	2.6④	27.8④	69.6④
墨西哥 Mexico	3.8	25.9	70.2
印度尼西亚 Indonesia	14.0	40.7	45.3
马来西亚 Malaysia	9.5④	50.4④	40.1④
泰　国 Thailand	9.6	46.9	43.5
新加坡 Singapore	0.1	33.8	66.1
韩　国 Korea.Rep.	3.7④	40.8④	55.5④
印　度③ India	18.6	27.6	53.8
巴　西 Brazil	9.8	37.9	52.3
俄罗斯联邦 Russian Federation	5.6	38.0	56.4

注：①发达国家指高收入国家，发展中国家指中低收入国家。②2001年数据。
③为2003年数据。④2004年数据。

资料来源：世界银行数据库；国际货币基金组织数据库；联合国数据库；联合国粮农组织数据库；联合国贸发会议《世界投资报告》2006年；联合国开发计划署《人文发展报告》2006年。

Note:①Developed countries refer to those with high revenue and developing countries those with low and medium revenue.②The data for 2001.③The data for 2003.④The data for 2004.

Sources:World Bank Database;IMF Database;UN Database;FAO Database;UNCTAD *World Investment Report,* (2006);UNDP *Human Development Report* (2006).

11-6 世界主要沿海国家（地区）就业人数

The Employment of the Major Coastal Countries(Regions)

单位：万人 Unit:10 000 persons

国家和地区 Country and Region	1990	2000	2003	2004	2005
中　国 China	64 749.0	720 850.0	74 432.0	75 200.0	75 825.0
中国香港 Hong Kong,China	271.2	320.7	321.9	330.9	338.6
孟加拉国 Bangladesh	5 015.9	5 176.4	4 432.2		
印　度① India		36 896.6			
印度尼西亚 Indonesia	7 585.1	8 983.8	9 078.5	9 372.2	9 494.8
以色列 Israel	149.2	222.1	233.0	240.1	249.4
日　本 Japan	6 249.0	6 446.0	6 316.0	6 329.0	6 356.0
哈萨克斯坦 Kazakhstan			698.5	718.2	
韩　国 Korea Rep.	1 808.5	2 115.6	2 213.9	2 255.7	2 285.6
马来西亚② Malaysia	668.5	932.2	987.0	998.7	1 054.8
巴基斯坦④ Pakistan	2 979.7	3 684.7	3 985.2	4 200.9	4 291.6
菲律宾 Philippines	2 253.2	2 777.5	3 155.3	3 174.1	3 287.5
新加坡 Singapore	153.7	209.5			226.7
斯里兰卡④ Sri Lanka	596.3	630.8	694.3	730.5	
泰　国⑤ Thailand	3 084.2	3 300.1	3 467.7	3 571.2	3 630.2
土耳其 Turkey	1 994.6	2 158.1	2 114.7	2 179.1	2 204.7
越　南 Viet Nam	302.9	3 836.8	4 117.6	4 231.6	
埃　及② Egypt	1 436.1	1 720.3	1 811.9		
南　非⑥ South Africa		1 171.2	1 162.2	1 151.8	
加拿大 Canada	1 316.5	1 475.9	1 566.5	1 595.0	1 617.0
墨西哥⑦ Mexico		3 820.3	3 950.0	4 097.6	4 132.1
美　国③ United States	11 879.3	13 520.8	13 773.6	13 925.2	14 173.0
阿根廷④ Argentina	437.3	826.2	895.6	941.5	963.9
巴　西④ Brazil	6 210.0		8 016.3	8 459.6	
委内瑞拉 Venezuela	640.4	882.2		1 003.6	
保加利亚 Bulgaria	409.7	298.0	316.7	323.6	327.6
法　国 France		2 326.1	2 463.1	2 472.0	
德　国 Germany		3 660.4	3 617.2	3 565.9	3 656.6
意大利 Italy	2 145.4	2 122.5	2 213.3		2 262.1
荷　兰② Netherlands	635.6	773.3	783.0	778.2	778.4
波　兰 Poland		1 452.6	1 361.7	1 379.5	1 411.6
罗马尼亚 Romania		1 076.4	922.3	915.8	914.7
俄罗斯联邦⑧ Russian Fed.		6 507.0	6 643.2	6 727.5	6 816.9
西班牙 Spain	1 257.9	1 550.6	1 729.6	1 797.1	1 897.3
乌克兰 Ukraine		2 017.5	2 016.3	2 029.6	2 068.0
英　国③ United Kingdom	2 693.5	2 779.3	2 782.1	2 800.8	2 816.6
澳大利亚 Australia	783.7	895.1	945.9	963.6	995.7
新西兰 New Zealand	148.1	177.9	192.1	201.7	207.3

注：①指在10人及10人以上公共企业和私营非农企业的雇员人数。②15至64岁。③16岁及以上。④10岁及以上。⑤13岁及以上。⑥15岁至65岁。⑦12岁及以上。⑧15岁至72岁。

资料来源：国际劳工组织劳工统计数据库。

Note:①refers to the number of employees in the public businesses and private,non-agriculturel businesses with ten or more persons.②15-64 in age.③16 years old and above.④10 and more in age. ⑤13 and more.⑥15-65 in age.⑦12 and above. ⑧15-72 in age

Sources:ILO Labour Statistics Database.

11-7 主要沿海国家(地区)渔业增加值
Fishery Added Value of Major Coastal Countries(Regions)

单位:亿本币 Unit: National Currency 100 million

国家和地区 Country and Region	渔业增加值 Fishery Added Value				
	2001	2002	2003	2004	2005
中国香港 Hongkong,China	10	10	8	9	8
孟加拉国 Bangladesh	1 390	1 426	1 478	1 548	
印 度 India	2 484	2 709	2 703		
印度尼西亚 Indonesia	37	47			
伊朗 Iran	2	2	3		
哈萨克斯坦 Kazakhstan	54	61	46		
韩 国 Korea Rep.	20 672	18 117	19 720	18 110	17 296
马来西亚 Malaysia	55	50	48	53	
菲律宾 Philippines	848	903	956	1 107	
新加坡 Singapore	2	2	2	2	1
泰 国 Thailand	1 110	1 061	1 040		
越南 Viet Nam	18	20	24		
加拿大 Canada	14	14			
墨西哥 Mexico	68	69	82	84	
阿根廷 Argentina	7	14	16	13	
委内瑞拉 Venezuela			1		
法 国 France	16	16	17	15	14
德 国 Germany	2	2	2	2	
意大利 Italy	13	14	15	15	17
荷兰 Netherlands	3	3	3	2	
波 兰 Poland	2	2	2	2	
西班牙 Spain	15	16	15	16	
乌克兰 Ukraine	1	1	1		
英 国 United Kingdom	4	4	4	3	
新西兰 New Zealand	3				

11-8 主要沿海国家(地区)旅馆和饭店业增加值

Added Value of Hotel Industry in the Major Coastal Countries(Regions)

单位:亿本币 Unit: National Currency 100 million

国家和地区 Country and Region	旅馆和饭店业增加值 Added Value of Hotel Industry				
	2001	2002	2003	2004	2005
中国香港 Hong Kong,China	331	302	258	331	
中国澳门 Macao,China	27	30	30	42	
孟加拉国 Bangladesh	174	194	220	250	
印 度 India	2 147	2 292	2 610		
伊朗 Iran	3	3	4		
哈萨克斯坦 Kazakhstan	189	258	390		
韩 国 Korea.Rep.	161 834	179 548	183 095	184 206	188 995
马来西亚 Malaysia	84	87	89	99	
菲律宾 Philippines	686	754	792	894	
新加坡 Singapore	35	34	31	35	
泰 国 Thailand	2 845	3 047	2 946		
越南 Viet Nam	15	17	19		
埃 及 Egypt	73	83	122		
加拿大 Canada	242	254			
墨西哥 Mexico	2 702	2 784	2 868	3 067	
美 国 United States	2 658	2 791	2 925	3 128	
阿根廷 Argentina	73	67	82	99	
委内瑞拉 Venezuela	2	2	2		
保加利亚 Bulgaria	5	6	6	7	
法 国 France	310	327	337	346	352
德 国 Germany	309	312	305	307	
意大利 Italy	438	435	442	467	488
荷兰 Netherlands	79	84	83	82	
波 兰 Poland	84	82	83	85	
罗马尼亚 Romania	25	32	40		
西班牙 Spain	461	498	526	575	
乌克兰 Ukraine	12	13	15		
英 国 United Kingdom	295	312	326	338	
澳大利亚 Australia	159	162	176	196	
新西兰 New Zealand	23				

11-9 主要沿海国家(地区)运输、仓储和通讯业增加值

Added Values of Transportation,Storage as Communications Industries in the Major Coastal Countries(Regions)

单位:亿本币 Unit: National Currency 100 million

国家和地区 Country and Region	运输、仓储和通讯业增加值 Added Values of Transportation,Storage and Commnuications Industries				
	2001	2002	2003	2004	2005
中国 China	8 406	9 393	10 098	12 148	10 526
中国香港 Hong Kong,China	1 175	1 218	1 174	1 268	1 438
中国澳门 Macao,China	28	30	27	31	45
孟加拉国 Bangladesh	2 552	3 111	3 444	3 810	4 047
印度 India	15 731	17 342	20 208		27 495
印度尼西亚 Indonesia	76	97	119	142	181
伊朗 Iran	56	66	83	100	134
以色列 Israel	326	326	340	352	377
日本 Japan	35	35	34	35	35
哈萨克斯坦 Kazakhstan	3 626	4 378	5 708	6 770	8 880
韩国 Korea Rep.	411 905	451 338	477 870	509 690	520 555
马来西亚 Malaysia	236	251	269	298	321
巴基斯坦 Pakistan	5 130	5 428	6 099	6 998	9 022
菲律宾 Philippines	2 476	2 769	3 132	3 668	3 912
新加坡 Singapore	176	178	173	195	231
斯里兰卡 Sri Lanka	1 587	1 765	2 009	2 359	2 728
泰国 Thailand	4 207	4 435	4 583	4 872	5 199
越南 Viet Nam	19	21	23	30	37
埃及 Egypt	311	342	405	446	507
南非 South Africa	895	1 000	1 104	1 223	1 325
加拿大 Canada	747	765	813	858	924
墨西哥 Mexico	5 951	6 116	6 457	7 255	7 883
美国 United States	6 402	6 480	6 665	7 126	9 381
阿根廷 Argentina	229	231	300	374	442
巴西 Brazil	575	635	783	818	933
委内瑞拉 Venezuela	6	7	8	13	17
保加利亚 Bulgaria	34	40	42	46	50
法国 France	822	893	912	945	976
德国 Germany	1 054	1 094	1 088	1 114	
意大利 Italy	844	909	919	957	994
荷兰 Netherlands	284	305	315	325	
波兰 Poland	481	533	555	591	619
罗马尼亚 Romania	117	147	182		149
俄罗斯联邦 Russian Fed.	7 252	9 056	11 249	16 708	19 436
西班牙 Spain	454	494	523	555	
乌克兰 Ukraine	246	275	319	427	510
英国 United Kingdom	716	744	783	783	
澳大利亚 Australia	526	564	596	647	669
新西兰 New Zealand	87	94	101	107	111

11-10 主要沿海国家（地区）渔获量及人均鱼类消费量

Catches as Per Capita Fish Consumption in the Major Coastal Countries(Regions)

国家和地区 Country and Region	海洋渔区标称渔获量（吨） Nominal Catch in the Marine Fishing Zones(ton)		人均鱼类消费量（克／天） Per Capita Fish Consumption(g/d)
	2004	2005	2004
世界总计 World Total	**85 788 203**	**83 718 042**	
阿尔巴尼亚 Albania	2 055	2 065	
阿尔及利亚 Algeria	140 000 *	126 259	
安哥拉 Angola	230 005	230 000	
安提瓜和巴布达 Antigua Barb	2 527	2 999	
阿根廷 Argentina	916 412	896 074	
澳大利亚 Australia	227 875	245 734	40.86
巴哈马 Bahamas	11 347	11 347	
巴林 Bahrain	14 259	11 854	
孟加拉国 Bangladesh	455 207	474 597	28.56
巴巴多斯 Barbados	2 500 *	1 869	
比利时 Belgium	26 239	24 071	
伯利兹 Belize	2 907	3 915	
贝宁 Benin	11 788	9 835	
巴西 Brazil	500 116	506 848	15.47
文莱 Brunei Darsm	2 417	2 390	
保加利亚 Bulgaria	5 818	3 409	7.77
柬埔寨 Cambodia	55 817	60 000	
喀麦隆 Cameroon	53 000 *	67 345	
加拿大 Canada	1 135 286	1 048 588	41.15
佛得角群岛 Cape Verde	8 446	7 742	
智利 Chile	4 935 376	4 330 325	
哥伦比亚 Colombia	100 801	92 000	
科摩罗群岛 Comoros	14 935	15 070	
刚果 Congo Rep.	17 737	25 868	
库克群岛 Cook Is.	3 279	3 737	
哥斯达黎加 Costa Rica	19 817	21 340	
科特迪瓦 Cote d'Ivoire	49 542	50 000	
古巴 Cuba	34 760	28 149	
塞浦路斯 Cyprus	1 527	1 886	
丹麦 Denmark	1 089 731	910 538	
多米尼加共和国 Dominican Rp.	12 243	9 412	
厄瓜多尔 Ecuador	335 411	407 323	
埃及 Egypt	111 466	107 453	38.54
萨尔瓦多 El Salvador	40 210	39 063	

续表 continued

国家和地区 Country and Region	海洋渔区标称渔获量（吨） Nominal Catch in the Marine Fishing Zones(ton)		人均鱼类消费量（克/天） Per Capita Fish Consumption(g/d)
	2004	2005	2004
赤道几内亚 Eq. Guinea	2 500 *	2 500	
爱沙尼亚 Estonia	85 533	96 300	
法罗群岛 Faeroe Is.	599 386	565 260	
福克兰群岛 Falkland Is.	55 368	84 545	
斐济 Fiji	43 642	37 200	
芬兰 Finland	100 316	95 476	
法国 France	594 898	572 248	52.92
波利尼西亚 Fr. Polynesia	12 147 *	12 102	
加蓬 Gabon	37 667	34 163	
冈比亚 Gambia	28 923	29 500	
格鲁吉亚 Georgia	2 900	2 948	
德国 Germany	239 492	264 268	32.34
加纳 Ghana	324 370	317 274	
希腊 Greece	91 280	91 158	
格陵兰岛 Greenland	216 302	216 302	
格林纳达 Grenada	2 039	2 050	
瓜德罗普 Guadeloupe	10 100	10 100	
危地马拉 Guatemala	6 531 *	4 948	
几内亚 Guinea	88 550	92 571	
几内亚比绍 Guinea Bissau	6 050 *	6 050	
圭亚那 Guyana	55 917	52 572	
海地 Haiti	8 000	8 000	
洪都拉斯 Honduras	14 839 *	19 100	
冰岛 Iceland	1 727 879	1 660 812	
印度 India	2 810 668	2 849 280	11.79
印度尼西亚 Indonesia	4 501 070	4 049 640	50.77
伊朗 Iran	314 165	343 492	11.56
伊拉克 Iraq	2 355	459	
爱尔兰 Ireland	279 619	262 026	
以色列 Israel	2 303	2 755	56.67
意大利 Italy	281 985	294 550	44.18
牙买加 Jamaica	13 071	12 969	
日本 Japan	4 341 326 *	4 018 937	114.29
肯尼亚 Kenya	7 774	7 104	
基里巴斯 Kiribati	31 600 *	34 000	
朝鲜 Korea D. P. Rep.	200 000 *	200 000	
韩国 Korea Rep.	1 565 035	1 631 569	116.33

续表 continued

国家和地区 Country and Region	海洋渔区标称渔获量（吨） Nominal Catch in the Marine Fishing Zones(ton)		人均鱼类消费量（克/天） Per Capita Fish Consumption(g/d)
	2004	2005	2004
拉脱维亚 Latvia	125 031	150 262	
黎巴嫩 Lebanon	3 601	3 523	
利比里亚 Liberia	6 359	6 000	
利比亚 Libya	46 073 *	46 073	
立陶宛 Lithuania	156 374	138 238	
马达加斯加 Madagascar	98 958	106 400	
马来西亚 Malaysia	1 331 645	1 209 601	127.97
马尔代夫 Maldives	158 576	185 980	
马耳他 Malta	1 134	1 435	
马尔提尼克 Martinique	6 200	5 500	
毛里塔尼亚 Mauritania	194 380	242 577	
毛里求斯 Mauritius	10 227	10 048	
墨西哥 Mexico	1 352 848	1 204 949	23.72
密克罗尼西亚 Micronesia	29 229	29 331	
摩洛哥 Morocco	893 758	931 814	
莫桑比克 Mozambique	25 924 *	29 466	
缅甸 Myanmar	1 132 340	1 204 286	51.10
纳米比亚 Namibia	569 208	551 195	
荷兰 Netherlands	519 536	547 108	60.66
荷属安的列斯群岛 Neth Antilles	600 *	2 422	
新喀里多尼亚 New Caledonia	3 500 *	3 315	
新西兰 New Zealand	538 786	534 582	51.25
尼加拉瓜 Nicaragua	18 431	29 567	
尼日利亚 Nigeria	282 987	285 131	18.59
挪威 Norway	2 521 820	2 392 427	
阿曼 Oman	165 029	150 571	
巴基斯坦 Pakistan	386 098	339 829	5.69
帛琉群岛 Palau Is.	1 079	932	
巴拿马 Panama	192 079	211 787	
巴布亚新几内亚 Papua New Guinea	220 300 *	236 780	
秘鲁 Peru	9 575 654	9 348 292	
菲律宾 Philippines	2 071 123	2 104 117	69.40
波兰 Poland	171 779	135 778	26.61
葡萄牙 Portugal	221 425	211 751	
波多黎各 Puerto Rico	2 428	2 551	
卡塔尔 Qatar	11 134	13 935	
留尼汪岛 Reunion	3 373	4 596	

续表 continued

国家和地区 Country and Region	海洋渔区标称渔获量（吨） Nominal Catch in the Marine Fishing Zones(ton)		人均鱼类消费量（克／天） Per Capita Fish Consumption(g/d)
	2004	2005	2004
俄罗斯 Russian Fed.	2 763 130	2 971 709	49.46
萨摩亚群岛 Samoa Is.	4 718	4 500	
圣多美和普林西比 Sao Tome & Principe	4 141	3 600	
沙特阿拉伯 Saudi Arabia	55 418	60 403	
塞内加尔 Senegal	395 263	355 070	
塞舌尔 Seychelles	93 740	106 555	
塞拉利昂 Sierra Leone	120 440	131 993	
新加坡 Singapore	2 173	1 920	
索罗门群岛 Solomon Is.	36 563 *	28 520	
索马里 Somalia	27 300 *	29 800	
南非 South Africa	881 039	816 708	18.30
西班牙 Spain	794 626	840 093	82.19
斯里兰卡 Sri Lanka	253 080	130 400	59.71
苏丹 Sudan	5 000 *	5 500	
苏里南 Suriname	32 577	39 724	
瑞典 Sweden	268 529	254 942	
叙利亚 Syria	3 077	3 677	
坦桑尼亚 Tanzania	49 270	49 300	
泰国 Thailand	2 645 488	2 396 987	64.08
多哥 Togo	23 013	22 732	
汤加 Tonga	1 673	1 900	
特立尼达和多巴哥 Trinidad & Tobago	9 709	13 414	
突尼斯 Tunisia	109 223	107 893	
土耳其 Turkey	504 897	380 381	19.18
特克斯和凯科斯群岛 Turks & Caicos	5 679	5 491	
乌克兰 Ukraine	197 413	238 916	42.31
阿拉伯联合酋长国 United Arab Em.	90 000 *	90 000	
英国 United Kingdom	651 285	668 286	45.10
美国 United States	4 930 704	4 863 916	34.78
乌拉圭 Uruguay	120 659	123 776	
瓦努阿图 Vanuatu	94 787	151 079	
委内瑞拉 Venezuela	440 910 *	420 910	38.97
越南 Viet Nam	1 733 434	1 791 100	39.73
也门 Yemen	256 300	263 000	

注：为联合国粮农组织估算值 (表11-11同)。

资料来源：《渔业统计年鉴》，联合国粮农组织，2005年Vol.96 (表11-11同)。

Note: It is estimated by FAO from available sources of information or calculation.(Same as Table 11-11).

Source: *Yearbook of Fishery*, FAO, Vol. 96, 2005 (Same as Table 11-11).

11-11 主要沿海国家（地区）海藻和其他养殖品种捕捞产量

Capture Production of Seaweeds and Other Aquatic Plants in the Major Coastal Countries(Regions)

单位：吨 Unit: ton

国家和地区 Country and Region	2003	2004	2005
世界总计 World Total	**1 273 948**	**1 432 612**	**1 305 803**
阿根廷 Argentina	…	…	
澳大利亚 Australia	8 879	12 570	14 167
加拿大 Canada	20 057	16 996	20 756
智利 Chile	309 055	390 577	409 851
爱沙尼亚 Estonia	1 126	…	809
斐济 Fiji	406	358	350
法国 France	69 866	64 723	76 633
冰岛 Iceland	22 119	21 470	21 096
印度尼西亚 Indonesia	64 610	70 440	7 730
意大利 Italy	2 000 *	1 850 *	1 600
日本 Japan	111 467	113 836	104 893
韩国 Korea Rep.	5 154 *	9 060	15 212
马达加斯加 Madagascar	1 699	…	…
墨西哥 Mexico	28 996	27 418	27 000
摩洛哥 Morocco	11 131	13 126	12 813
纳米比亚 Namibia	408	…	…
挪威 Norway	153 209	148 322	153 906
秘鲁 Peru	7 864	7 418	5 000
菲律宾 Philippines	721 *	433	298
葡萄牙 Portugal	375	253	624
俄罗斯 Russian Fed.	39 080	58 024	50 262
塞内加尔 Senegal			
南非 South Africa	32 000 *	30 000 *	6 619
西班牙 Spain	1	525	441
坦桑尼亚 Tanzania	2 500 *	240	240
乌克兰 Ukraine	233	255	522
美国 United States	49 496	35 339	35 922

11-12 主要沿海国家（地区）油气可开采储量(2005年)

Recoverable Reserves of Oil and Gas in the Major Coastal Countries(Regions),2005

国家和地区 Country and Region	原油和液化天然气（万吨） Crude Oil and liquified Natural Gas(10 000 tons)	天然气（亿立方米） Natural Gas (100 million m^3)	油页岩(万吨) Oil Shale(10 000 tons)
中国① China	330 000	17 110	
孟加拉国 Bangladesh	800	2 880	
印 度 India	74 000	6 880	
印度尼西亚 Indonesia	70 500	20 510	
伊朗 Iran	1 264 900	231 790	
以色列 Israel	100	10	60 000
日 本 Japan	900	320	
马来西亚 Malaysia	39 300	22 720	
缅 甸 Myanmar	700	3 110	
巴基斯坦 Pakistan	3 900	5 950	
菲律宾 Philippines	600	800	
泰 国 Thailand	7 900	1 880	170 000
土耳其 Turkey	3 900	120	2 000
越南 Viet Nam	33 800	1 450	
埃 及 Egypt	49 200	7 350	
尼日利亚 Nigeria	425 200	34 760	
南非 South Africa	500	230	1 300
加拿大 Canada	80 200	20 960	
墨西哥 Mexico	233 700	2 490	
美 国 United States	380 100	51 980	
阿根廷 Argentina	37 600	6 700	
巴 西 Brazil	134 800	2 250	
委内瑞拉 Venezuela	1 113 900	41 820	
保加利亚 Bulgaria	200	50	
法 国 France	2 100	230	
德 国 Germany	6 000	3 850	
意大利 Italy	8 900	2 450	
荷兰 Netherlands	1 500	17 610	
波 兰 Poland	1 600	1 170	
罗马尼亚 Romania	10 000	4 470	
俄罗斯联邦 Russian Fed.	821 900	447 620	
西班牙 Spain	200	20	
英 国 United Kingdom	59 100	13 310	
澳大利亚 Australia	28 100	13 660	365 100
新西兰 New Zealand	900	680	

注：①联合国统计数据。

资料来源：联合国《能源统计年鉴》2003年。

Note:①Data from United Nations.

Source:United Nations *Energy Statistics Yearbook* 2003.

11-13 主要沿海国家(地区)天然气产量
Natural Gas Production of the Major Coastal Countries(Regions)

单位:万亿焦耳 Unit:terajoule

国家和地区 Country and Region	天然气产量 Production of Natural Gas	
	1990	2005
世界总计 World Total	**74 772 000**	**104 215 000** ②
中国①China		1 965 624
孟加拉国 Bangladesh		434 844 ③
印 度 India	389 000	1 074 840 ③
印度尼西亚 Indonesia	2 345 000	3 204 684
哈萨克斯坦 Kazakhstan		203 700 ③
马来西亚 Malaysia	502 000	2 032 680 ④
巴基斯坦 Pakistan		897 336 ③
泰 国 Thailand	226 000	924 348
加拿大 Canada	4 114 000	6 566 448
墨西哥 Mexico	955 000	2 245 812
美 国 United States	18 255 000	19 931 436
阿根廷 Argentina	793 000	2 141 976 ④
巴 西 Brazil	151 000	742 356 ④
乌克兰 Ukraine		684 780
丹麦 Denmark	122 000	437 592
法 国 France	77 000	33 228
德 国 Germany		661 068
匈牙利 Hungary	16 000	108 744
爱尔兰 Ireland	88 000	20 880
意大利 Italy	653 000	458 004
荷兰 Netherlands	2 541 000	2 624 388
挪威 Norway		3 527 904
波 兰 Poland	100 000	179 748
罗马尼亚 Romania	959 000	491 508
俄罗斯联邦 Russian Fed.		24 785 724
乌克兰 Ukraine		684 780
英 国 United Kingdom	1 904 000	3 693 300
澳大利亚 Australia	797 000	1 113 360
新西兰 New Zealand	180 000	156 864

注：①联合国统计数据。②2002年数据。③2003年数据。④2004年数据。

Note:①Data from the United Nations.②Data for 2002.

③Data for 2003.④Data for 2004.

11-14 主要沿海国家(地区)能源利用效率

Use Efficiency of Energy Resources in the Major Coastal Countries(Regions)

单位:国际元/千克标准油 Unit:PPP$ per kg of oil equivalent

国家和地区 Country and Region	单位能源生产的GDP GDP per unit energy production		
	2001	2002	2003
世界 World	**4.65**	**4.70**	**4.72**
中国[①] China	4.73	4.75	4.54
中国香港 Hong Kong,China	10.57	10.89	10.94
孟加拉国 Bangladesh	10.20	10.26	10.36
印度 India	4.93	4.97	5.23
印度尼西亚 Indonesia	4.13	4.13	4.26
伊朗 Iran	3.08	3.20	3.22
以色列 Israel	7.32	7.21	7.08
日本 Japan	6.38	6.36	6.48
哈萨克斯坦 Kazakhstan	1.68	1.78	1.86
韩国 Korea Rep.	4.06	4.41	4.22
马来西亚 Malaysia	3.97	4.03	3.88
巴基斯坦 Pakistan	4.17	4.25	4.23
菲律宾 Philippines	7.28	7.50	7.84
新加坡 Singapore	4.00	4.05	4.54
斯里兰卡 Sri Lanka	8.29	8.59	8.82
泰国 Thailand	5.05	4.98	5.02
土耳其 Turkey	5.78	5.90	5.97
越南 Viet Nam	4.28	4.26	4.40
埃及 Egypt	5.13	5.13	5.15
尼日利亚 Nigeria	1.13	1.17	1.29
南非 South Africa	3.91	4.03	3.88
加拿大 Canada	3.45	3.54	3.44
墨西哥 Mexico	5.78	5.65	5.55
美国 United States	4.29	4.35	4.51
阿根廷 Argentina	7.32	7.17	7.22
巴西 Brazil	6.96	6.93	6.86
委内瑞拉 Venezuela	2.47	2.37	2.26
保加利亚 Bulgaria	2.57	2.74	2.84
法国 France	5.81	5.94	5.86
德国 Germany	6.01	6.15	6.10
意大利 Italy	8.46	8.49	8.16
荷兰 Netherlands	6.00	6.03	5.81
波兰 Poland	4.51	4.62	4.60
罗马尼亚 Romania	3.75	3.92	4.02
俄罗斯联邦 Russian Fed.	1.73	1.84	1.93
西班牙 Spain	7.08	7.05	7.02
乌克兰 Ukraine	1.67	1.73	1.88
英国 United Kingdom	6.81	7.04	7.09
澳大利亚 Australia	4.67	4.67	4.81
新西兰 New Zealand	4.40	4.53	4.82

注：①世界银行统计数据。

Note:①Data from World Bank.

11-15 主要沿海国家(地区)国际海运装货量和卸货量

International Ocean Shipping Loading and Unloading Capacities in the Major Coastal Countries(Regions)

单位：万吨 Unit:10 000 tons

国家和地区 Country and Region	国际海运装货量 International Ocean Shipping Loading Capacity		国际海运卸货量 International Ocean Shipping Unloading Capacity	
	2000	2004	2000	2004
中国① China	146 400 ②	180 000 ②		
中国香港 Hong Kong,China	6 770	8 603	10 693	13 486
中国澳门 Macao,China	8 ③	6 ③	16 ③	13 ③
孟加拉国 Bangladesh	89	76	1 408	1 568
印度尼西亚 Indonesia	14 153	24 608	4 504	8 225
伊朗 Iran	3 065	3 328	4 486	5 798
以色列 Israel	1 576 ④	1 592 ④	2 222 ④	3 347 ④
日本 Japan	13 010	17 849	80 654	82 907
韩国 Korea Rep.	15 078	23 495	41 882	50 869
马来西亚 Malaysia	5 483 ⑤	7 758 ⑤	6 922 ⑤	9 744 ⑤
缅甸 Myanmar	306	451 ⑥	467	582 ⑥
巴基斯坦 Pakistan	617	1 060	3 080	3 650
新加坡 Singapore	32 618 ②	39 300 ②		
土耳其 Turkey	1 652		1 836	
埃及 Egypt		1 273		2 851
南非 South Africa	13 361	12 848 ⑥	2 598	4 184 ⑥
美国 United States	34 334	34 963	83 352	95 624
阿根廷 Argentina			1 550 ⑦	1 632 ⑦
法国 France	6 810 ⑧	9 610 ⑧	20 273 ⑧	22 358 ⑧
德国 Germany	8 602	9 970	14 725	16 420
荷兰 Netherlands	9 940	11 279	32 507	35 106
波兰 Poland	3 152	3 894	1 582	1 715
罗马尼亚 Romania	1 265	1 823	1 283	2 231
俄罗斯联邦 Russian Fed.	828	821	84	113
西班牙 Spain	5 627	7 556	19 343	23 653
乌克兰 Ukraine	4 271	6 542	684	1 140
澳大利亚 Australia	48 750	58 697	5 418	6 527
新西兰 New Zealand	2 214	2 236	1 379	1 849

注：①联合国统计数据。②包括装货量和卸货量。③仅指集装箱运输。④不包括石油。
⑤包括沿岸运输、转载。数据仅指马来西亚半岛。⑥2003年数字。⑦不包括再出口。⑧包括沿岸运输。

Note：① The data from the United Nations Database.②Including both goods loaded and unloaded.③Containerized cargoes only.④Excluding petroleum.⑤Including transshipments.Data refer to Peninsular Malaysia only. ⑥Data for 2003.⑦Encluding re-exports,excluding packing.⑧Including coastwise traffic.

11-16 世界主要外贸海运量（2005年）

World Major Maritime Freight Traffic of Foreign Trade, 2005

品 种 Sort	海运量（百万吨） Freight Traffic (million tons)		海运周转量（十亿吨海里） Ton-Kilometer (billion ton-sea mile)	
	2005	所占比例（%） Percentage	2005*	所占比例（%） Percentage
合 计 Total	**6 784**	**100.0**	**28 868**	**100.0**
原 油 Crude Oil	1 820	26.8	8 985	31.1
成品油 Refined Oil	488	7.2	2 475	8.6
铁矿石 Ironstone	650	9.6	3 801	13.2
煤 炭 Coal	690	10.2	3 091	10.7
谷 物 Corn	242	3.6	1 384	4.8
其 他 Others	2 894	42.7	9 132	31.6

注：为估计数。
资料来源:海运统计与市场评论。（表11-19同）
Note:Estimated data.
Source: Shipping Statistics and Market Review,ISL (same as Table 11-19).

11-17 主要沿海国家（地区）国际旅游人数
Number of International Tourists of Major Coastal Countries(Regions)

单位：万人 Unit:10 000 persons

国家和地区 Country and Region	国外游客到达人数 Arrivies Number of Foreign Tourists			出国旅游人数 Number of Tourists Going Abroad		
	2000	2003	2004	2000	2003	2004
世界总计 World Total	**69 296**	**70 625**	**77 592**	**80 934**	**81 257**	**85 427**
高收入国家 High Income	44 658	44 161	47 010	39 747	39 808	42 924
中等收入国家 Middle Income	22 003	23 855	27 561	28 911	29 075	29 473
低收入国家 Low Income	1 536	1 761				
中国 China	3 124	3 297	4 176	1 047*	2 022*	2 885*
中国香港 Hong Kong,China	1 305	1 553	2 181	461	442	500
孟加拉国 Bangladesh	19	24	27	112	141	
印度 India	264	272		441	535	
印度尼西亚 Indonesia	506	446	532			
伊朗 Iran	134	154	165	228		
以色列 Israel	241	106	150	353	329	361
日本 Japan	475	521	613	1 781	1 329	1 683
哈萨克斯坦 Kazakhstan	147	241	307	124	237	391
韩国 Korea Rep.	532	475	581	550	708	882
马来西亚 Malaysia	1 022	1 057	1 570	3 053	3 220	3 076
缅甸 Myanmar	20	20	24			
巴基斯坦 Pakistan	55	50	64			
菲律宾 Philippines	199	190	229	167	180	
新加坡 Singapore	691	570		444	422	
斯里兰卡 Sri Lanka	40	50	56	52	56	68
泰国 Thailand	957	1 008	1 173	190	215	270
土耳其 Turkey	958	1 334	1 682	528	592	729
埃及 Egypt	511	574		296	364	
尼日利亚 Nigeria	81					
南非 South Africa	587	650	667	383		
加拿大 Canada	1 962	1 753	1 909	1 918	1 773	1 959
墨西哥 Mexico	2 064	1 866	2 061	1 107	1 104	1 249
美国 United States	5 123	4 121	4 608	6 132	5 625	6 177
阿根廷 Argentina	290	299	335	495	308	338
巴西 Brazil	531	413	479	322	229	
委内瑞拉 Venezuela	46	33	49	95	83	81
保加利亚 Bulgaria	278	404	463	233	340	388
法国 France	7 719	7 504	7 512	1 988	1 857	2 113
德国 Germany	1 898	1 839		7 440	7 460	
意大利 Italy	4 118	3 960	3 707	2 199	2 681	2 334
荷兰 Netherlands	1 000	918	964	1 389	1 646	
波兰 Poland	1 740	1 372	1 429	5 667	3 873	2 722
罗马尼亚 Romania	327			638	649	697
俄罗斯联邦 Russian Fed.	2 116	2 251	2 205	1 837	2 046	2 441
西班牙 Spain	4 789	5 183	5 359	410	409	512
乌克兰 Ukraine	643	1 251		1 342	1 479	
英国 United Kingdom	2 520	2 471	2 775	5 683	6 142	6 419
澳大利亚 Australia	493	474	521	349	338	
新西兰 New Zealand	178	210	233	128	137	173

注：*为国内居民出境总人数。

Note:* is numbers of domestic residents outbound.

11-18 集装箱吞吐量居世界前20位的港口（2006年）

The World First Twenty Seaports of Container Handled, 2006

单位：万TEU Unit: 10 000TEU

港 口 Seaport	所属国家或地区 Country or Territory	吞吐量 Cargo Handled
新加坡 Singapore	新加坡 Singapore	2 479
香港 Hongkong	中国 China	2 354
上海 Shanghai	中国 China	2 172
深圳 Shenzhen	中国 China	1 847
釜山 Pusan	韩国 Korea Rep.	1 204
高雄 Gaoxiong	中国台湾 Taiwan China	977
鹿特丹 Rotterdam	荷兰 Netherlands	969
汉堡 Hamburg	德国 Germany	886
迪拜 Dubayy	阿联酋 United Arab Em	878
洛杉矶 Los Angeles	美国 United States	847
青岛 Qingdao	中国 China	770
长滩 Long Beach	美国 United States	729
宁波舟山 Ningbo Zhoushan	中国 China	714
安特卫普 Antwerp	比利时 Belgium	702
广州 Guangzhou	中国 China	666
巴生港 Kelang	马来西亚 Malaysia	632
天津 Tianjin	中国 China	595
纽约/新泽西 New York and New Jersey	美国 United States	513
丹戎帕拉帕斯 Tanjung Periuk	马来西亚 Malaysia	477
不来梅 Bremen	德国 Germany	445

资料来源：《2006年中国航运报告》。

Source: *2006 China Shipping Report.*

11-19 货物吞吐量居世界前20位的港口（2005年）
The World First Twenty Seaports of Cargo Handled, 2005

单位：百万吨　Unit: million tons

港　口 Seaport	所属国家或地区 Country or Territory	吞吐量 Cargo Handled	备　注 Remark
上海 Shanghai	中国 China	443.2	外贸货物 foreign trade cargo
新加坡 Singapore	新加坡 Singapore	423.2	内外贸货物 domestic and foreign trade cargo
鹿特丹 Rotterdam	荷兰 Netherlands	370.2	内外贸货物 domestic and foreign trade cargo
宁波 Ningbo	中国 China	268.8	内外贸货物 domestic and foreign trade cargo
广州 Guangzhou	中国 China	250.4	外贸货物 foreign trade cargo
天津 Tianjin	中国 China	240.7	内外贸货物 domestic and foreign trade cargo
香港 Hongkong	中国 China	230.1	内外贸货物 domestic and foreign trade cargo
南路易斯安那 Southern Louisiana	美国 United States	220.4	内外贸货物 domestic and foreign trade cargo
釜山 Pusan	韩国 Korea Rep.	217.2	内外贸货物 domestic and foreign trade cargo
休斯敦 Houston	美国 United States	195.4	内外贸货物 domestic and foreign trade cargo
名古屋 Nagoya	日本 Japan	187.1	外贸货物 foreign trade cargo
青岛 Qingdao	中国 China	186.8	内外贸货物 domestic and foreign trade cargo
光阳 Gwang	韩国 Korea Rep.	177.5	内外贸货物 domestic and foreign trade cargo
大连 Dalian	中国 China	170.9	内外贸货物 domestic and foreign trade cargo
洛杉矶 Los Angeles	美国 United States	169.0	内外贸货物 domestic and foreign trade cargo
秦皇岛 Qinhuangdao	中国 China	169.0	内外贸货物 domestic and foreign trade cargo
千叶 Chiba	日本 Japan	165.7	内外贸货物 domestic and foreign trade cargo
长滩 Long Beach	美国 United States	160.1	内外贸货物 domestic and foreign trade cargo
安特卫普 Antwerp	比利时 Belgium	160.1	内外贸货物 domestic and foreign trade cargo
深圳 Shenzhen	中国 China	153.5	内外贸货物 domestic and foreign trade cargo

11-20 海上商船拥有量居世界前20位的国家或地区（2005年）

The World First Twenty Countries or Territories of Maritime Merchant Ships, 2005

国家和地区 Country and Region	艘数（艘） Number of Vessels	载重吨 Deadweight ton	
		万吨 10 000 tons	占世界% Percentage
世界总计 World Total	**28 375**	**88 576.2**	**100.0**
希腊 Greece	2 943	15 941.3	18.0
日本 Japan	3 009	13 101.0	14.8
德国Germany	2 731	7 027.9	7.9
中国 China	2 568	6 418.7	7.2
美国 United States	1 077	4 689.7	5.3
中国香港 Hongkong, China	661	4 338.2	4.9
挪威 Norway	1 287	4 329.2	4.9
韩国 Korea Rep.	918	2 942.1	3.3
中国台湾Taiwan,China	537	2 431.9	2.7
新加坡 Singapore	661	2 227.4	2.5
英国 United Kingdom	616	2 133.2	2.4
丹麦 Denmark	645	1 900.9	2.1
俄罗斯Russia	1 473	1 548.8	1.7
意大利 Italy	644	1 404.4	1.6
印度 India	299	1 355.3	1.5
沙特阿拉伯Saudi Arabia	98	1 132.2	1.3
瑞士Switzerland	354	1 126.8	1.3
比利时Belgium	149	1 105.7	1.2
土耳其Turkey	785	1 045.3	1.2
伊朗 Iran	156	978.5	1.1

注：1. 表中数据按载重吨排序。
2. 范围为1000总吨及以上船舶，截止日期均为2006年1月1日。
3. 因统计口径不同，表中数据与其他出版物公布的数据稍有差别。

Note: 1. The data on China in the table have not been verified, so they are provided just for reference.
2. The ships, ranging over 1000 tons and more in gross ton, are counted as of Jan. 1, 2006.
3. Owing to different statistical requirements, the data given in the table may be somewhat different from those issued in other publications.